LOW CARB
HIGH FAT
AND PALEO

Slow Cooking

LOW CARB HIGH FAT AND PALEO

Slow Cooking

Birgitta Höglund

60 Healthy and Delicious LCHF Recipes

Skyhorse Publishing

CONTENTS

Fish and Shellfish

Organ Meats

Stocks and Bouillons

Bread

Sauces and Tasty Green Side Dishes

Foreword by Ann Fernholm

"Daddy, Daddy—can I play with the tail?" The father quickly chops off the tail and throws it to his six-year-old child, who, along with a friend, happily starts playing with it. They whisk the red dirt, trailing the tail's end like a brush. The father continues to remove the hide and cut up the cow. His work must be done quickly—it's hot outside and the meat will start to go bad quickly. He cuts off the branded ear and tosses it into the bush. No one will be any the wiser as to who owned this dead animal. It's the neighbor's cow, which strayed onto the Hayfield Shenandoah Cattle station land. This means the meat is free, even though more than 20,000 head of cattle graze on this land. The workers eat beefsteak for breakfast.

This memory from the wide expanses of Northern Australia comes to me as I stand in the meat department of a food market in Stockholm (Sweden) some thirty years later. The assistant has just pulled out a packet of oxtail from the freezer. The tail is cut into eight pieces; it's very clear that this is a tail, because the pieces become progressively thinner as they move away from the base. I chuckle. In my freezer you'll see mostly chicken breasts and salmon filets. I have never bought oxtail. The meat is marbled with fat, and in the middle is the tailbone. I think to myself: this is the beginning of something new, and walk home. These are the first steps toward undertaking something I've promised myself—namely, to explore new flavors.

Cut Down on Sugar and Discover a Whole New World

All the recipes within this book contain very few carbohydrates, and Birgitta Höglund doesn't skimp on the fat. Since I began eating according to low-carb guidelines, I have taken an interest in noticing how my taste for different foods has changed. After only about a year, I'm not as tempted by sweet rolls and cookies as I used to be—they simply don't taste that great anymore. They're too sweet. Things I used to love to eat have now become sort of underwhelming. On the flip side, other tastes have taken over; spices have somehow blossomed in magnitude and diversity.

Becoming aware that fatty food has little bearing on weight gain or on the increased risk of heart attack has gone a long way toward heightening my enjoyment of cooking. Before, each calorie led me to suffering a guilty conscience. Now I enjoy each mouthful of food, and feel good about it. Food is far more flavorful when I don't have to be sparing in my use of fat, so I willingly trade sweet rolls for cream sauces. This knowledge has brought on a renewed enthusiasm for cooking and spices; consequently, I was more than happy to help out when I was asked to write the foreword to this cookbook. This is the starting point to a new adventure: to learn how to slow-cook. I want to discover new foods, new spices, and new ingredients—such as oxtail.

Slow cooking requires sturdy pots, however—not your everyday cookware. My husband found a cast-iron Dutch oven from Skeppshult's, the famous old Swedish foundry, in a thrift store, for about $53. It was in this pot that I browned the onions and garlic, the first step to Birgitta Höglund's Oxtail Stifado.

Once the tomato purée began to simmer with the onion, the first heady aromas wafted up from the pot. The orange, cinnamon, cloves, and bay leaves followed—a beguiling treat for the nose. I browned the chunks of oxtail in another pan, before combining everything in the cast-iron pot. All the spices simmered and blended together while cooking in the oven over a period of eight hours. The oven's temperature: 212°F (100°C).

That evening, there was a certain French quality to the air as I dished out the finished stew: its dark color, tender meat, and flavorful seasonings all made it look like a Christmassy bœuf bourguignon. I asked my children to describe their first impressions. "Gingerbread, but with real ginger," answered my six-year-old son. My daughter, who enjoyed the stew immensely, picked up on the subtle notes of orange, and the cinnamon. Oxtail Stifado became even more interesting when I mixed the stew with some of the recommended mint-flavored yogurt. The flavors became more nuanced and landed gently on the tongue: first the orange, then the clove and cinnamon, and lastly, the full-bodied flavor of the meat. Paired with some red wine, it became a meal to appease the soul.

Fat—The Newly Discovered Sixth Taste Area on our Tongues

We might be able to attribute some of the feeling of calm we experience after eating tasty and fatty foods to newly discovered receptors on our taste buds that react specifically to fat. We learned in biology class that our tongue is able to distinguish between four basic tastes: sweet, salty, sour, and bitter. Later, scientists found that we are also capable of tasting the amino acid glutamate, which is a part of protein-rich foods. This taste is called umami, from the Japanese word for good taste.

This is nowhere near the end of the story on our perception of taste, however. Fat hasn't got a flavor per se, which is why scientists think it might be erroneous to call those new receptors on the tongue 'taste' buds. They make us very aware that it is fat we're tasting in the food— they're akin to a small recorder. When we put something fatty in our mouth, the receptors ensure that the nerve signal is sent to the brain. Scientific research shows that both dopamine and endorphins are released; these neurotransmitters make our body feel calm and relaxed.

When I'm enjoying a bowl of Oxtail Stifado, I'm sure this is what happens. The endorphins travel out into the bloodstream—I can feel it. Suddenly I realize that as I'm eating, I'm slowing down. The body is beginning to feel that it has had enough food.

Fat Makes You Feel Satiated for Extended Periods of Time

By this time the oxtail has been absorbed in the stomach, and several other hormones are on their way out into the bloodstream. There are several molecules that influence the point at which we feel full: leptin, ghrelin, peptid YY, glucagon-like peptide-1, and cholecystokinin. You don't need to try to remember all these; I simply mention them to illustrate the fact that the body takes feelings of satiety very seriously.

Hormones cross over between the stomach, the digestive tract, adipose tissue, muscles, and the brain. All this ensures that we benefit from all the nourishing aspects of food in the most efficient way possible.

One of the hormones related to satiety deserves a little closer attention, especially if you think about how we've been advised to eat as little fat as possible. This hormone is called oleoylethanolamide; researchers have shortened it to OEA. This is how it works: we have taste receptors in our digestive tract that are identical to those that recognize fat on our tongue. The sensors in our digestive tract serve a different function, however: when fat attaches itself to the sensors in the digestive tract, the hormone OEA is released. This satiety hormone is therefore specifically linked to fat, and its special characteristic is to prolong our feeling of satiety. Mice suddenly eat less and lose weight when scientists inject them with OEA; they seem satisfied with their food for longer periods of time.

I was often hungry shortly after eating lunch when I was following a low-fat diet. During my coffee break in the afternoon, I'd break down and eat two or three cookies. My stomach was grumbling—almost in pain—on my way home from work. I had to hurry to prepare something to eat.

When eating a low-carbohydrate diet, hunger becomes a nonissue for long stretches of time. I can't say that I feel full all of the time, but my mind isn't as preoccupied with feelings of hunger and thoughts of food. Many others who follow a low-carb regimen have experienced the same effect. One explanation could be that more OEA is released in the body when we eat more fat. Limiting carbohydrates also helps to keep blood sugar levels on a low and even keel. In Sweden today, approximately one million people suffer from metabolic syndrome. It's a disruption of the body's metabolism that instigates spikes in blood sugar and alters blood fats. One way metabolic syndrome manifests itself is in the form of abdominal obesity, but many people of normal weight also suffer from metabolic syndrome. Elevated blood sugar levels bring on inflammation in the body; researchers call this low-grade inflammation. As the immune system is constantly at work, this brings on feelings of illness and fatigue, especially in the afternoon. Many suffer from aches and pains, and depression can readily set in.

We've believed for a long time now that the only way to deal with this is to cut calories and to go on restrictive diets. Scientists have thought—perhaps logically—that the amount of energy we consume correlates with how much body fat we accumulate. It's not until recently that they've started to test the validity of this assumption using scientific methods. What science is now showing is that people who are told to cut down on carbohydrates naturally eat just enough, and often lose weight too. Perhaps this is because, among other things, fatty food releases more OEA in our digestive tract. In the 1970s, when the Swedish National Food Agency advised us to cut down on our consumption of dietary fat, there was still so much we didn't know about the human body. At the time, nobody knew that feelings of satiety are directly influenced by the different kinds of food we eat.

It's such a relief to follow a low-carbohydrate lifestyle. It's wonderful not to have to count calories. It's a new and refreshing experience to be able to eat and not feel guilty about it. For Birgitta Höglund, the low-carbohydrate diet has been even more life changing: she is free of inflammation and body pain, and she has become far more energetic. That's great, because it has given her the stamina to write cookbooks full of delicious recipes that will tempt others to go out on culinary adventures. If you're trying out a low-carbohydrate diet for the first time, don't concentrate solely on losing weight or lowering your blood sugar, as that will only make you focus on all the sweets you can't eat anymore. Instead, think of all the new food you're going to try. In low-fat cooking, spices never get the chance to develop fully. Now, with fat allowed back on your plate, you can experience full flavors once again without the slightest twinge of anxiety.

Good food is central to happiness. I'm just on my way out to visit the meat market to buy an elk roast and a pumpkin. The cast-iron pan is going to work its magic again this weekend.

My New Life with Natural Foods

It may be the case that, in today's increasingly stressful world, we don't pay enough attention to how our food choices affect our general well-being. At least, that was my situation up until about five years ago. These days, I think about what I eat on a daily basis, or I should say, about what I don't eat—notably gluten and sugar.

I've experienced some truly significant health improvements as a result of changes that I've made in my daily diet. My story begins in 2000, when I injured my back while on the job as a chef. The chronic pain I suffered after a spinal fusion surgery, and later on from the onset of fibromyalgia, is now almost entirely gone after five years of gluten-free, low-carbohydrate living. The fatigue I used to feel after eating "regular" food full of carbohydrates (mostly from grains) is a distant memory, too.

The day after attending a lecture by Annika Dahlqvist, MD, in which she spoke on how she cured herself of fibromyalgia, I decided to try her nutritional model, which is LCHF (Low Carb High Fat). This was the beginning of a thrilling journey toward my new, healthy life.

Prior to this, during long periods of sick days, food preparation in my home was kept very simple: it often consisted of spaghetti, covering most of my plate. To go with this, a small chunk of bologna or ham, chopped onion, and a splash of cream, topped off with lashings of ketchup—often the lone "vegetable" of the meal. Sandwiches weren't much better, featuring lots of bread and a few fixings such as cheese or ham, with maybe a slice of cucumber on top.

Carbohydrates from pasta, sandwiches, and buns have now been replaced with lots of tasty, richly colored vegetables, most of them grown above ground. In addition to quantities of healthy fiber, vegetables and herbs contain many nutrients such as vitamin C and other essential vitamins, minerals, and trace elements—the tiny things we need in order to feel good.

My intake of protein from meat, fish, shellfish, and eggs increased significantly once I altered my diet. I finally reached the level of protein (18 to 20 percent of daily calories) that the Swedish National Food Agency recommends. Many people still believe that those of us who follow LCHF/Paleo diets consume more meat and protein than is endorsed by the official dinner plate model, but that is not true. Protein is necessary to build muscles and to stay healthy, and we should take care to not eat it in excess.

I've always liked fat, so my portions of pasta and sandwiches were often topped with a thick layer of butter. Low-fat products have never had a place in my kitchen—I've always cooked with natural fats. I'm proud of my many years spent as a chef at various restaurants, even if, at times, I did not have the strength to prepare more than the simplest meals for myself after my back injury.

The combination of simple carbohydrates with a lot of butter is, as most of us are aware today, absolutely disastrous for weight management. It was no different for me—my weight crept up slowly but surely to a high of somewhere around 210 pounds (95 kg); I stopped weighing myself after this eye-opening discovery.

When I changed my diet to natural, low-carbohydrate foods after attending Dr. Dahlqvist's lecture, I discovered that I could replace large amounts of carbohydrates with fat from butter, coconut fat, olive oil, heavy cream, and cheese.

Naturally, this made food far more appetizing, while at the same time keeping me full and content for hours. By contrast, my past way of eating made me exhausted due to abrupt swings in blood sugar that usually followed carbohydrate-rich meals. Now I avoid the blood sugar rollercoaster entirely with my new diet, and my blood sugar stays level throughout the day.

If you're a newcomer to low-carbohydrate nutrition, you might be wondering why I have combined LCHF and paleo in the book's title. The reason is that the differences between the two nutritional models are negligible. LCHF includes dairy products, but excludes most fruit, nuts, and root vegetables due to their high carbohydrate content.

Although followers of paleo/Stone Age diets don't eat dairy, many do opt to eat butter. Their nutritional guidelines allow for more liberal amounts of fruit, nuts, berries, and root vegetables. Both nutritional models are basically free of gluten and sugar, substances that are precisely the culprits responsible for body pain and inflammation, which in turn seem to be precursors to many common ailments we suffer from in the West.

Over the years, I have turned toward a more paleo-style diet, but I still eat a lot of butter, a small amount of goat's and sheep's cheese, aged cheese, and occasionally some full-fat milk products. My stomach has been the main beneficiary of this diet; I live a more or less pain-free existence since I changed my lifestyle. Many who suffer from chronic pain or digestive trouble notice an improvement in their health when they start eating natural, low-carbohydrate foods, and abstain from ingesting gluten or milk protein.

As Ann Fernholm so vividly and engagingly describes in the foreword of this book, the flavors of the dishes I prepare are deeper and more complex, thanks to the increased amount of fat they contain.

Freshly cooked broccoli with butter and salt flakes; creamy vegetable gratin; fresh salad with herbs and a good dressing—these are but a few of the new side dishes on my menu, replacing pasta, potatoes, and rice.

Today, my protein comes straight from the source; I seldom eat any meat that has traveled by way of a grocery store's meat or deli counter. The food in my refrigerator is most often produced or raised organically, because the taste, quality, and nutritional content simply can't be beat. I also keep in mind that food animals have a better quality of life when they're able to roam outside and graze on what they were always meant to eat. Organic vegetables are far more flavorful than those that are conventionally grown, too; we're also able to steer clear of pesticide residue.

There are several advantages to using slow cooking when preparing these quality foods. Low temperatures ensure that both flavor and nutrition aren't leached out from ingredients; tastes deepen and intensify. Also, we can cook cuts of meat that are typically too tough to quickly pan-fry or sear. This is good for our pocketbooks and the environment.

Slow-cooked dishes can take a long time to prep, but, paradoxically, they're also time savers. How can this be? Making a meal from scratch does take a bit longer than simply heating up a manufactured, ready-to-eat meal.

Those of us who eat food with fewer carbohydrates really don't need to eat as often, so a lot of kitchen time is saved that way. Most people who follow LCHF/paleo diets are fully satisfied by eating breakfast, lunch, and dinner; some of us only eat twice a day because our food is so nutritious. No time is wasted on planning snacks or nighttime meals. Another reason slow cooking saves time is because you need not stay and stir anything by the stove—the cooking pretty much takes care of itself.

Once you've added all the ingredients to the pot or slid them into the oven, things can proceed almost without supervision. If you own an electric Crock-Pot or other slow cooker, you can let it stand and simmer the entire day while you're at work or out taking care of other things. The same applies for a Dutch oven with a tight-fitting lid, set in the oven on low heat.

You can also leave a clay pot in the oven for a few hours. What could be more perfect than bringing the pot straight from the oven to the table upon your return from the gym or a run on the ski slopes?

The recipes in this book are prepared according to my new lifestyle's philosophy. For me, LCHF/paleo is not a temporary diet but the way I have chosen to eat for the rest of my life. I hope this book will lead you to a better and more flavorful life, too.

Eat, enjoy, and feel great!
Birgitta Höglund

Tips for Successful Slow Cooking

Slow-cooked food has always had a place in my life. I grew up in the countryside in the northeastern, mountainous region of the Swedish province of Dalarna, and slow-cooked elk stew was frequently served in my childhood home. The elk my father hunted was often prepared the traditional way: the bones were simmered for many hours to give up all the goodness in their marrow for the stock, which together with the tender meat became the base for dishes like dill stew, elk collops, and meat soup.

The elk blood was turned into a delicious blood pudding, which was cooked in an enameled milk container, standing in a water bath (*bain-marie*) on the stove. My recipe for blood pudding in this book is gluten free but tastes almost the same as the one my mother used to make.

My mother was exceptionally clever at cooking something tasty from every edible part of the elk. Even the skull was used—the tongue was cooked together with the Christmas ham. The tongue is delectable with a sauce made from morels. The heart and cheeks were ground and used in her tasty meat hash, but today those organ meats are considered a delicacy, just like beef cheek. Elk liver can taste potent to many, but it becomes mild and delicate after standing overnight in the pantry, soaking in a milk bath.

To this day, I still have a deep appreciation for this type of cooking. My interest in food, which began during childhood, blossomed and later became my profession as I worked to become a chef. Throughout my many years in different restaurant kitchens, I've uncovered even more secrets of cooking long and slow.

Big baking trays filled with chopped veal bones, vegetables, herbs, and spices were left in the oven to brown. The bones were simmered for two days before being strained out of the liquid; the stock was then reduced to concentrated meat glaze. It was kept in a large bucket in cold storage—a layer of fat on top to keep it from spoiling—and was used to season all the sauces, stews, and soups.

Lately, slow cooking has enjoyed a welcome renaissance. More and more of us take pride in our heirloom cuisine. During the Stone Age, cooking meant wrapping food in leaves and burying it in the ground with heated cooking stones. Many hours later the meat came out tender and juicy without getting scorched.

Later, clay was added to enclose the leaves. The risk of burning food decreased further still, and the meat's juices remained under the clay cover, making the food even more flavorful and nutritious.

The same cooking method developed throughout many cultures; these creations became juicy meat stews left overnight in the after-heat of bakery and restaurant ovens. You can still find local restaurants in the Mediterranean where they will serve meat stews prepared this way.

This method has worked very well, even in Sweden. Over the past few years, I've been involved in developing recipes for the restaurant PriMaten (a play on the word Primate) in the university town of Uppsala. This restaurant always has an oven going on very low heat overnight, which is used for preparing different slow-cooked dishes, which are ready to serve as soon as the first lunch guests arrive the next day.

Clay pots were something I used a lot in the 1970s when they were a popular trend. I have since bought a new clay pot that will certainly see a lot of use for many years to come.

Over the last few years, I have even tested electric slow cookers, also known generically as Crock-Pots. Crock-Pot and Nordic Cooker are popular brands in Sweden. Their automatic timers simplify the cooking process, because they allow most of the cooking to go unattended. As very little liquid is needed in these pots, the end result is usually very flavorful.

I like the feel and heft of a cast-iron pan for everyday cooking. It's important to read the accompanying user manual carefully first if you use a clay pot or an electric Crock-Pot to achieve the best results. Below, you'll find a description of different cooking methods I use at home in my own kitchen. Unless otherwise noted, all my recipes in this book serve 4.

Pot or Dutch Oven on the Stovetop

In my opinion, cast-iron pots work best for slow cooking. Both enameled and pure cast-iron pots disperse heat evenly to the cooked food. If you cook your stew or roast on the stovetop, you'll need to add more liquid to prevent the meat from drying out, as the heat on the stove is hotter than in the oven.

Simmer over low heat, with the lid on, and add in more liquid as it reduces—some homemade stock, for example, or stock made from a bouillon cube mixed with warm water.

Cast-Iron Pot in the Oven

When cooking food in the oven, it's important to use a heavy cast-iron pot such as a Dutch oven with a tight-fitting lid, as this will distribute the heat more evenly and minimize the risk of the food drying out. The lower the heat you set the oven to, the less likely you will be to burn the dish. If the pot is left at 167°F–176°F (70°C–80°C) it can be left in the oven for a whole day, or eight to ten hours overnight. It's a very easy and convenient way to cook your food if you want to brown bag your lunch the next day.

If you're in more of a hurry, bump up the temperature to 212°F (100°C), and test after four to six hours to see if the meat is cooked—doneness will depend on the size and type of meat you are cooking. This stew might also need to have some liquid added to it during the cooking time.

Clay Pot in the Oven

Meat cooked in a clay pot is very tender and flavorful. The meat's surface is also seared in a way that can't be replicated in a cast-iron pot. It's important to soak the pot and the lid in water for about fifteen minutes before putting them to use.

A clay pot is always placed, lid on, in a cold oven; not until it is in the oven should the oven be turned on. I usually set the oven to 347°F (175°C) to minimize the risk of burning the food. The dish typically takes three to four hours to cook, depending on what I'm making. Whole roasts, chicken, and stews made with cheaper cuts such as chuck roast or oxtail take a long time but turn out fantastically tender. Ground meat and vegetable dishes are ready after only about an hour. I suggest you use a dedicated fish clay pot when cooking fish dishes, as the odor of fish is notoriously difficult to get rid of when washing the dishes. Fish dishes will be ready in about thirty to forty-five minutes.

Very hot steam is created by this cooking method, so be extra careful when removing the lid, and always remove it away from you. If you want more of a seared surface on your food, remove the lid toward the end of the cooking time.

Electric Slow Cooker and Crock-Pot

Today's slow cookers—also called Crock-Pots—have simplified cooking for those who don't like to spend a lot of time in the kitchen. The Crock-Pot doesn't need constant watching; the food can stand and simmer the entire day while you're at work, and it's ready to eat when you get home. Energy consumption is also far less than for traditional cooking on the stove or in the oven.

Crock-Pots don't require as much liquid as when you're preparing a stew on the stovetop, so use less stock than my recipes call for if you choose to use an electric slow cooker.

When cooking with a Crock-Pot, you can mix all of the ingredients together in a bowl, and pour them all into the pot without having to sear or sauté any of them first. Personally, I think the flavor of the dish is much enhanced if the meat and vegetables have been seared first.

This means that I'll take a bit of extra time to do this before turning the food over to the Crock-Pot, which then does most of the cooking, largely unattended.

Cook on the low power setting for six to ten hours; the time will vary a little, depending on what brand of slow cooker you're using, so take the time to carefully study the cooker's user manual before you put it to use.

How to Tie Up a Roast

A roast holds an even, consistent shape and slices easily if you tie it up with kitchen twine before cooking it. Make a loop with the twine, put it around one end of the roast and pull it tightly—much like you would when wrapping a gift.

Pull the twine about an inch farther along the length of the roast; place your thumb on top and make another loop around the width of the roast. Bring the twine through the loop, pull it tightly and move the twine along the roast a little further, and repeat with a new loop. Pull the twine tightly, and repeat to make loops evenly along the entire length of the meat. Continue until your reach the end of the roast.

Pull the twine across the end of the roast, turn the roast over, and carry the twine around each diagonal piece of string. Cut off the twine and fasten it where you started tying it.

If you find it challenging to tie the roast evenly with one long piece of twine, simply cut shorter lengths and tie them individually across the roast. Once the roast is done cooking and has had a chance to rest and retain its juices, cut off the twine and cut the roast into even slices.

Which Fat is Best for Cooking?

I prefer to use butter for frying, but you can use organic coconut oil or fat if you prefer your cooking to be free of dairy. I fry in a neutral, unflavored coconut oil, except in dishes where the coconut taste is actually called for.

Cold-pressed virgin olive oil adds great flavor to Mediterranean dishes. I prefer to mix olive oil with butter when frying—this makes it is easier to gauge the correct temperature for frying.

Thank Yous

I would like to take this opportunity to thank all the shops in the Swedish town of Östersund, which so kindly lent me their products to help me make this book as beautiful as possible. Many thanks go to Cervera, Village, Nyströms Kakel, and Kulturarvet.

My book's mouthwatering dishes were made using superior produce from local food producers in the Swedish northern province of Jämtland whenever possible. Everything was purchased at Ost & Vilt (Cheese & Game) and Kvantum.

The attractive cookware comes from Acreto/Crock-Pot, Nordic Cooker, and Le Creuset.

With his camera, Mikael Eriksson has, yet again, captured the beautiful soul in my food.

Last but not least, many thanks go to my mother Ingegerd, who taught me much about the bygone art of slow cooking.

BEEF

Oxtail Stifado

When I spend time in Greece, I like to eat slow-cooked stews made with lamb, pork, and rabbit, as well as other dishes made from tougher cuts of meat left to simmer a long time to bring out their wonderful, tender texture and rich flavor.

Stifado, a stew featuring hints of cinnamon and orange, is a staple of most local restaurants. They usually make it with stew meat; here I've used a nutritious cut—oxtail.

- 12 pearl onions
- 4 cloves garlic
- ¼ cup (50 ml) tomato purée
- 2.2 lbs. (1 kg) oxtail, cut into pieces
- 1–1 ½ tablespoons salt
- ½ tablespoon black peppercorns, coarsely crushed
- 1 organic orange
- 1 ¼ lbs. (500 g) tomatoes, crushed (from a can or a carton)
- 2 tablespoons red wine vinegar
- 2 cinnamon sticks
- 2 bay leaves
- 8 whole allspice berries
- 1 tablespoon oregano (Mediterranean, not Mexican)
- ¼ teaspoon ground cloves
- 1–1 ½ tablespoons salt
- about ½ tablespoon black peppercorns, coarsely crushed
- butter for frying

Peel the onions and garlic, and cut them in half. Brown them slightly; add in the tomato purée and let cook with the onions for a few minutes—this will remove some of the acidic tang from the purée. Transfer the onion mixture into a Crock-Pot or a stew pot.

Sear the oxtail in a sauté pan; season with salt and pepper. Wash and julienne the orange peel with a zester or a potato peeler. Juice the orange, and mix juice and orange peel into the stew along with the crushed tomatoes, vinegar, cinnamon, bay leaves, allspice, oregano, and ground cloves.

Place the meat into the sauce, and deglaze the sauté pan with 1 ¾ fl. oz. (50 ml) water; pour the resulting sauce over the stew. Let it simmer in the Crock-Pot on low for about 7 to 8 hours. On the stovetop it will take between 4 to 5 hours. Test and see if the meat is tender; if not, let it cook a little longer. Taste and adjust for seasoning as needed.

To cook the stifado in the oven, let the stew come to a boil, and then set it, covered with a tight-fitting lid, in an oven set at 212°F (100°C). Check on it after a few hours, turning the chunks of meat, and thin the sauce with some more water if it gets too thick.

After 8 hours, use a fork or toothpick to check and see if the meat is tender. If it's not ready, let it simmer for another hour.

Meat Sauce with Sun-Dried Tomatoes

Many cook their meat sauce in about twenty minutes when throwing together a quick meal. If you take the time to let it simmer for two hours, however, you'll end up with a dish that's heavenly. The ground meat becomes so tender that it melts in your mouth, and all the spices will have fully bloomed.

For a milder variation of this sauce, stir in a cup of full-fat cream, and let it simmer for quarter of an hour at the end of the cooking time. If you prefer a sauce with more full-bodied, robust taste, use a full-bodied red wine instead of water.

- 2 yellow onions
- 2 cloves garlic
- 1 carrot
- 1 ¾ oz. (50 g) hot-smoked pork belly
- 1 lb. 5 oz. (600 g) ground beef
- 1 ¾ oz. (50 g) butter
- ¼ cup (50 ml) tomato purée
- 2 tablespoons balsamic vinegar
- 2 tablespoons whole pink peppercorns, crushed
- 1 tablespoon oregano (Mediterranean)
- 1 tablespoon paprika
- 1 tablespoon Worcestershire sauce
- about 1 teaspoon salt
- about ½ teaspoon coarsely ground black pepper
- 8 sun-dried tomatoes
- 1 organic meat stock cube + 6 ¾ fl. oz. (200 ml) water OR 6 ¾ fl. oz. (200 ml) homemade veal stock (see recipe on p. 107)

Finely chop the onion and garlic. Cut the carrot and pork belly into small cubes; sauté in half of the butter until the onion becomes translucent, and transfer to a stew pot. Brown the ground beef in the remaining butter; fry for a few minutes while stirring constantly, making sure all the lumps are gone. Add in the tomato purée and let it cook for a short while to soften the acidity of the puree.

Season this mix with vinegar, pink peppercorns, oregano, paprika, Worcestershire sauce, salt, and pepper; pour the meat mixture over the vegetables and combine thoroughly. Add stock to the pot and bring to a boil.

Let the sauce simmer over low heat for 2 hours, stirring from time to time. Dilute with water if the sauce looks too thick.

I make the accompanying white cabbage "pasta" by grating white cabbage coarsely, parboiling it in boiling, salted water, and then frying it in butter until soft, taking care not to let it brown. Season it lightly with salt and pepper.

Lasagna with Meat Sauce and Kale

Lasagna is one of the few pasta dishes I occasionally miss. Here I've layered a slow-cooked, flavorful meat sauce with parboiled kale instead of lasagna noodles.

The cheese sauce thickens by itself, without the addition of flour. It contains quark (fresh cheese), grated cheese, heavy cream, and eggs, which makes the lasagna very nutritious.

Lasagna

- 14 oz. (400 g) kale
- meat sauce or ground lamb stew (see the recipes on p. 22 and 52)

Cheese Sauce

- 4 organic eggs
- 8 ¾ oz. (250 g) quark (10 percent fat)
- 1 ¼ cups (300 ml) aged cheese, grated
- 3 ⅓ fl. oz. (100 ml) heavy cream
- about ½ teaspoon salt
- about ¼ teaspoon white pepper
- ¼ teaspoon nutmeg

Preheat the oven to 350°F (175°C). Butter an ovenproof dish.

Cut off the thicker stalks of the kale. Dump the kale in boiling, salted water and let it simmer for about 5–10 minutes, depending on the thickness of the leaves. Transfer the kale to a colander, and rinse it with cold water to stop the cooking. Drain well.

Stir together all the ingredients for the cheese sauce and mix thoroughly. Set a layer of kale at the bottom of the prepared dish. Add layers of meat sauce and cheese sauce, just like for regular noodle lasagna. Top with a layer of cheese sauce.

Bake in the middle of the oven for about 45 minutes; cover with foil if the lasagna begins to darken. Let the lasagna cool a little in the dish before serving to make slicing it easier. The lasagna is also tasty when made with parboiled white cabbage.

Goulash Soup

Twenty years ago I was chef for two winter seasons at a small mountain ("fjeld") hotel, with a view over the Swedish ski paradise in Åre valley. My goulash soup was a huge hit with the hungry skiers.

The old recipe contained diced potato, but now I use mushrooms and tasty vegetables that grow above ground instead.

- 1 ¼ lbs. (500 g) chuck roast, boneless
- 2 yellow onions
- 3 tomatoes
- 1 green bell pepper
- 1 yellow bell pepper
- 1 red bell pepper
- 3–4 cloves garlic
- 8 ¾ oz. (250 g) mushrooms
- about ½ tablespoon salt
- about 1 teaspoon white pepper
- 1 tablespoon crushed caraway seeds
- 1 tablespoon paprika
- 2 tablespoons tomato purée
- 1 ¾ oz. (50 g) butter
- 2 cups (500 ml) chicken stock (see p. 106) OR 2 organic chicken stock cubes + 2 cups (500 ml) water
- 6 ¾ fl. oz. (200 ml) dry red wine
- 1 tablespoon Worcestershire sauce
- 1 tablespoon dried parsley
- 1 tablespoon dried thyme
- 2 bay leaves

Cut the meat in small chunks, about ½ in. x ½ in. (1 cm x 1 cm). Cut onions, tomatoes, and bell peppers into small cubes. Press the garlic. Cut in half and finely slice the mushrooms.

Cook the meat in a stew pot until nicely browned. Add in the onion and garlic; season with salt, pepper, caraway seeds, paprika, and tomato purée. Let it cook a little more to let the spice flavors develop fully.

In a separate pan, sauté the mushrooms until golden; transfer them to the pot with the meat, and add in the stock, wine, Worcestershire sauce, herbs, and bay leaves.

Bring to a boil and let simmer on low heat for 2 to 3 hours, until the meat is tender. Stir now and then, diluting with more water if the goulash starts to looks too thick. Taste for seasoning. Serve it with a dollop of crème fraîche and some dinner muffins (recipe is on p. 118).

Slow-Cooked Beef Shank with Salami

Slices of beef shank with some bone marrow in the middle make a very flavorful and nutritious stew. Here we've been inspired by the Italian kitchen and have added in salami, white wine, and sage. A tangy gremolata is a lovely garnish.

- 2 teaspoons paprika
- about 2 teaspoons salt
- about ½ teaspoon red chili flakes
- 4 slices of beef shank, about 1 ¾ lbs. (800 g)
- 4 shallots
- 2 cloves garlic
- 2 green bell peppers
- 3 ½ oz. (100 g) salami, sliced
- 1 ¾ oz. (50 g) butter
- 6 ¾ fl. oz. (200 ml) dry white wine
- 1 ¾ fl. oz. (50 ml) veal stock (see p. 107) + 3 ⅓ fl. oz. (100 ml) water OR 1 organic meat stock cube + 5 fl. oz. (150 ml) water
- ¼ cup (50 ml) sage, coarsely chopped

Gremolata

- 2 cloves garlic
- 1 organic lemon
- ⅓ cup + 1½ tablespoons (100 ml) finely chopped Italian parsley

Gremolata: You can make this ahead. Peel and mince the garlic. Wash the lemon and grate the peel finely, chop into small pieces, and mix with the parsley.

Preheat the oven to 260°F (125°C).

Mix the paprika, salt, and chili flakes, and season the meat all over, letting it rest to absorb the spices. Cut a few slits in the fatty layer to prevent the surface of the meat from bulging while cooking. Peel and cut the shallots and garlic into thin wedges. Julienne the bell peppers and salami.

Sear the meat in half the butter on both sides until golden brown; transfer it to a stew pot. Brown the onion, bell peppers, and salami in the remaining butter. Pour the vegetables over the meat. Add in the wine, stock, and sage, and bring to a boil.

Position the pot in the lower part of the oven, and cook it for approximately 4 to 5 hours. The stew takes only slightly less time if cooked on the stovetop.

Remove the stew pot from the oven and check to see if the meat is tender; if it isn't ready, put it back in the oven to cook for another hour. Serve the meat with the gremolata, and some cold-pressed olive oil drizzled over.

If you cook the shanks in an electric Crock-Pot, place the vegetables at the bottom of the vessel and decrease the amount of wine and water to about 3 ⅓ fl. oz. (100 ml). With the Crock-Pot set on low, the cooking time will be about 6 to 8 hours. The timing will depend on what kind of slow cooker you use, so check the meat after 6 hours.

Old-Fashioned Jellied Veal (Head Cheese)

Jellied veal, or veal brawn, is an example of traditional homey fare that is eminently suitable for a modern low-carbohydrate lifestyle, as brawn contains practically no carbohydrate at all. Veal bone marrow is highly nutritious, and with slow cooking, all the nutrients are released into the stock.

- 2.2 lbs. (1 kg) veal with bone
- 4 ¼ cups (1000 ml) water + 1 tablespoon salt
- 1 yellow onion
- 1 small carrot
- 8 white peppercorns
- 5 allspice berries
- 4 whole cloves
- 2 bay leaves
- 3 sheets of gelatin (2 ¼ teaspoons in powder)
- 1 tablespoon acetic acid (about 12 percent)
- ½ teaspoon white pepper
- some added salt, if needed

Rinse the meat under cold water, and put it in a saucepan. Measure the water into a measuring cup, and check to see how much is needed to JUST cover the meat. Add 1 tablespoon of salt per quart of water. Bring the water to a boil and pour it over the meat. Bring it to back to a boil, and skim off the foam thoroughly with a slotted spoon. Cut the onion and carrot into small chunks, and add them to the meat along with the spices. Bring to a boil; lower the heat, cover with a lid, and let everything simmer for about 1 ½ to 2 hours, or until the meat is tender.

Remove the meat from the pot with a slotted spoon, let it cool a little, and then pull the meat from the bones. Remove fat and gristle, and put them back in the stock along with the bones. Boil the stock vigorously for about 30 minutes to allow the bones to release all their gelatin and the stock to thicken. Pour the stock through a sieve over a bowl to remove the bones, vegetables, and spices.

Soak the gelatin (or follow instructions on a packet of gelatin powder) in cold water while the bones are cooking, and cut the meat into very small bits. Or, if you prefer, grind the meat coarsely, or pulse it quickly in a food processor.

Measure the amount of meat in a measuring cup and add it to the equivalent amount of stock in a saucepan. Bring it to boil and mix in the gelatin, acetic acid, and white pepper. Taste and adjust for salt.

Mix well, let it cool a little, and then pour the jelly into molds that have been rinsed with cold water. Once they're cold, let them sit in the refrigerator overnight. Loosen the sides of the brawn by drawing a knife along the sides of the molds. If the jelly doesn't loosen, dip the mold quickly in warm water. Overturn the brawn onto a platter. The jelly can be frozen in the molds. Once defrosted, melt the brawn in a saucepan, and pour it into a mold to reset it.

Serve the brawn with red cabbage salad (recipe on p. 140).

Chuck Roast in Tomato Sauce

Chuck is a very flavorful cut of meat that requires a long period of cooking to become tender. The meat is from the muscular front section of the animal, where it's quite marbled with fat. This recipe features larger pieces of boneless chuck that are slow cooked in a robust tomato sauce.

Serves 5 to 6

- 8 ¾ oz. (250 g) pearl onions
- 3 large cloves garlic
- 1 can cherry tomatoes, peeled
- 5 fl. oz. (150 ml) dry red wine
- 2 tablespoons Dijon mustard
- 1 tablespoon dried thyme
- 1 teaspoon finely crushed caraway
- grated peel from one organic lemon
- 2 bay leaves
- 2.2 lbs. (1 kg) chuck, boneless
- butter for frying
- about 1 tablespoon salt
- about 1 teaspoon white pepper
- 6 ¾ fl. oz. (200 ml) veal stock (see p. 170) OR 1 organic meat stock cube + 6 ¾ fl. oz. (200 ml) water
- olive oil for drizzling

Preheat the oven to 212°F (100°C).

Set the pearl onions in warm water for a while to make them easier to peel. Peel onions and garlic; cut the garlic in half lengthwise, but leave the onions whole. Lightly brown the onions in a stew pot. Mix in the tomatoes, wine, mustard, thyme, and caraway. Bring to a boil.

Wash the lemon and finely grate the peel with a zester, or peel the outer yellow layer with a potato peeler, and finely julienne it with a knife. Add the lemon peel and the bay leaves to the stew pot.

Cut the meat into rectangular chunks. Sear them on both sides, doing this in several batches to avoid crowding the frying pan; season with salt and pepper. Add the pieces of meat to the tomato sauce once they're seared. Deglaze the pan with some stock.

Pour the stock over the meat in the stew pot, stirring so the meat is covered in sauce.

A little olive oil drizzled on top adds some extra flavor to the stew. Cover the pot with a tight-fitting lid and place it in a baking tin in the bottom section in the oven.

Let the stew cook for 4 to 5 hours. Check the meat with a toothpick to see if it's tender; if not, let it cook for another hour. A creamy avocado along with some sautéed zucchini or broccoli makes a very nice side dish to this stew.

Sunday Roast with Blueberry Cream Sauce

This pot roast's amazing flavor brings back taste memories from long ago. Served with my blueberry cream sauce, this is a luxurious meal that's also very easy to prepare.

- 2.2 lbs. (1 kg) beef knuckle
- about ½ tablespoon salt
- about 1 teaspoon white pepper
- 2 yellow onions
- 2 carrots
- butter, for frying
- 3 ⅓ fl. oz. (100 ml) veal stock (see p. 107) + 3 ⅓ fl. oz. (100 ml) water OR 6 ¾ fl. oz. water + ¼ organic meat stock cube
- 4 small canned Swedish sprat filets (or anchovies)
- 10 white peppercorns
- 5 allspice berries
- 1 bay leaf

Tie up the meat with kitchen twine (follow the description on how to do this on p. 16) if the meat isn't sold already bound in a net. In the stew pot, sear the roast all over, and season it with salt and pepper. Cut carrots and onions in chunks, and brown them with the roast, taking care not to let them get too brown.

Pour in the water; add the sprats or anchovies and the spices. Bring to a boil. Turn down the heat to low, place a lid on the stew pot, and let the roast simmer for about 1 ½ to 2 hours. Baste with the stock a few times. Turn the meat after 1 hour. Add more water if needed. Test the meat for tenderness with a fork.

Reduce the stock by boiling it for 10 minutes. Strain and reserve 1 ¾ fl. oz. (50 ml) of it for the cream sauce. Freeze the rest in an ice cube tray to have on hand for flavoring other tasty sauces.

This stovetop cooked roast can also be prepared in the oven, in a covered stew pot, at 260°F (125°C) for approximately 2 to 2 ½ hours.

Blueberry Cream Sauce

I also like using bilberries in this cream sauce, if you can find them. Blackberries are another option.

- 3 ⅓ fl. oz. (100 ml) reduced stock from the roast
- 6 ¾ fl. oz. (200 ml) blueberries
- 1 ¼ cups (300 ml) heavy cream
- ½ teaspoon balsamic vinegar
- salt and white pepper
- 1 oz. (25 g) butter, cut into small pieces

Bring the stock and the blueberries to a boil in a saucepan, and let them simmer until they form a thick jam. Add in the cream, vinegar, and spices, and let the sauce simmer for half an hour. Strain the sauce through some cheesecloth into another saucepan, and bring to a boil.

Remove the sauce from the heat and whisk in the butter, piece by piece. Taste the sauce and adjust for additional vinegar or spices.

Veal Stew in Dill Sauce

This is one of my favorite recipes among traditional Swedish homey fare. I make my dill stew using veal, but it is equally delicious with lamb. Here I've used a piece of boneless meat. If you choose to use bone-in meat you'll end up with a stock that's extra nutritious and delicious.

The sugar that's traditionally added to this dish is of course missing from my recipe. We derive some natural sweetness from onion, carrots, and heavy cream.

- 1 lb. 5 oz. (600 g) boneless veal shoulder
- about 2 teaspoons salt
- about 2 cups (500 ml) boiling water
- 1 yellow onion
- 2 carrots
- a few stalks of dill, taken from a fresh bunch (1 ¾ oz., or 50 g)
- 1 bay leaf
- 10 white peppercorns

Dill Sauce:
- 6 ¾ fl. oz. (200 ml) stock, from cooking the meat
- 1 ¼ cups (300 ml) heavy cream
- about ½ tablespoon acetic acid (12 percent)
- about ⅛ teaspoon white pepper
- salt
- 6 ¾ fl. oz. (200 ml) coarsely chopped dill

Rinse the meat well. Bring the water to a boil. Place the meat into a stew pot, sprinkle it with salt, and pour in boiling water to cover the meat. (Using boiling water means there will be fewer particles to skim off the surface of the stock.) Bring the water back to a boil, and skim off all the foam and particles with a slotted spoon.

Chop the onion and carrots in small chunks, and add them to the stew pot along with the dill stalks, bayleaf, and peppercorns.

Let the meat simmer on low heat for about 1 ½ to 2 hours, until it's tender.

Remove the meat from the pot with a slotted spoon, and reduce the stock by boiling it over high heat for about 5 minutes.

Strain the stock, and measure out the stock needed for the sauce. Freeze the remaining stock to keep on hand for flavoring another sauce or soup.

Bring the stock and the cream to a boil, and let cook until the sauce has reduced and thickened—this will take about half an hour. Season with acetic acid, white pepper, and more salt if needed. Put the meat back in the sauce for a few minutes to reheat it.

If you prefer a slightly thicker sauce, add in an egg yolk mixed with a few tablespoons of heavy cream. Stir this into the sauce once the meat has reheated, and let it simmer just until the sauce has thickened a little more. DO NOT BRING THE LIQUID TO A BOIL after the yolk has been added, or you'll end up with scrambled eggs.

Add in the chopped dill right before serving, as it loses flavor, color and nutrients if it is allowed to cook.

PORK

Pulled Pork Afelia

This dish is called pulled pork (the same as in Swedish) and it's very trendy at the moment. The name is derived from the fact that the meat is cooked for a long time over very low heat, and when it's ready to eat it's literally pulled apart by using a couple of forks. The meat is cooked so tender that it readily falls from the bone and shreds very easily into threads.

Most recipes I've seen for pulled pork use ready-made, bottled marinades that contain both sugar and MSG. My version of pulled pork is inspired by afelia, a stew made from pork marinated in red wine; it's a very nourishing, slow-cooked stew that can be found in traditional restaurants both in Greece and in Cyprus.

Serves 8 to 10

- 2 large red onions
- 1 head of garlic
- 6 ¾ fl. oz. (200 ml) dry red wine
- 3 ⅓ fl. oz. (100 ml) cold-pressed olive oil
- 2 tablespoons crushed coriander seeds
- 2 teaspoons dried thyme
- 2 teaspoons coarsely ground black pepper
- 2 teaspoons cinnamon
- 3.3 lbs. (1.5 kg) whole pork collar
- about 1 tablespoon salt

Peel and slice the red onions into thin wedges. Cut the garlic cloves in half. Mix all the ingredients for the marinade. Place a large freezer bag inside a larger freezer bag, and add half the onion mix to the bag. Rinse the pork collar, dry it well, and rub it all over with salt. Place the collar in the freezer bag and pour the marinade over it. Press out all air from the bag, seal the bag shut, and place it in a bowl. Leave the bowl in the refrigerator for at least 12 hours, preferably longer.

Preheat the oven to 260°F (125°C).

Place the meat and the marinade in an oven-safe casserole dish. Close it with a tight-fitting lid, and place the dish in the lower part of the oven for about 5 to 6 hours. In the picture, the meat was cooked in an electric slow cooker; it turned out extremely juicy and delicious. If using a Crock-Pot, the meat will be ready in about 8 to 12 hours if set on low, but it also depends on the brand of the slow cooker you're using.

To serve, pull the meat apart with two forks and mix it thoroughly with the gravy. Taste and adjust for salt. The meat is delicious on a slice of walnut bread, the recipe for which you'll find on p. 112. Other good sides are fennel slaw (see recipe on p. 138) and blue cheese butter (see recipe on p. 148).

Oven-Roasted Pork Belly with Orange and Chili

Fresh pork belly is great food for those of us who want to keep our food low in carbohydrates and free from all the additives typically found in brined pork and bacon.

This meat becomes very tender from slow cooking; its flavor stems from a tangy marinade with orange, and its spiciness from the chili.

- 1 lb. 5 oz. (600 g) fresh pork belly, with rind
- 1 organic orange
- 1 red chili pepper
- 1 tablespoon finely chopped rosemary
- 2 tablespoons gluten-free tamari soy sauce
- about 1 tablespoon salt, for the meat rub prior to roasting

Rinse the meat in cold water and dry it well. Slit the rind with a sharp knife at even intervals; only make surface slits in the rind—do not cut into the meat itself.

Wash the orange and cut off thin strips of peel with a zester. Or, peel off the outer layer with a potato peeler and cut it into fine strips with a knife.

Seed and finely chop the chili pepper. Caution: use food-service latex gloves to do this and make sure not to rub your eyes! Do not remove the seeds if you prefer more heat in the dish.

Strip the rosemary branch of its needles, and chop the needles finely. Juice the orange and mix it with the tamari, chili, rosemary, and orange peel.

Place the prepared meat in doubled plastic freezer bags. Pour the marinade over the meat and seal the bags shut. Work the marinade around the meat a little to make sure it's evenly distributed. Leave the plastic bags with the meat on a platter in refrigerator for 24 to 48 hours. Turn the meat a few times during this time.

Preheat the oven to 430°F (225°C).

Pour the marinade into a smaller ovenproof dish. Distribute the salt evenly on both sides of the pork collar, and rub it thoroughly into the cuts in the rind.

Place a baking rack inside the dish, and set the meat on the rack with the rind side up. Place the dish in the oven and let the meat brown for 20 minutes. Lower the temperature to 212°F (100°C) and let the meat roast for 2 hours. Baste it several times with the marinade.

Once it's cooked, let the meat rest for a while on the cutting board before cutting it into slices. Sautéed mushrooms and a Brussels sprout gratin covered in blue cheese (see recipe on p. 147) make excellent side dishes to this meat.

Thick-Cut Spareribs with Lime and Thyme

Thick-cut spareribs are highly marbled with fat, so they become extremely juicy after many hours in the oven.

Marinade

- 1 organic lime
- 1 bunch fresh thyme
- 2 tablespoons olive oil, cold pressed
- 2 tablespoons apple cider vinegar

For Roasting in the Oven

- 3.3 lbs. (1.5 kg) thick-cut spareribs, in one rack
- about 1 tablespoon salt
- about 1 teaspoon white pepper
- 1 carrot
- 4 red onions, small

Wash the lime and use a zester on the peel, or finely grate the peel. Juice the lime into a mortar and add the peel to the juice. Remove the thyme leaves from the sprigs and chop them coarsely. Pound them in the mortar along with the lime's juice. Add oil and vinegar and mix well.

Rinse the spareribs and dry them with paper towels; place the ribs in doubled plastic freezer bags. Pour the marinade over the ribs and seal the bags shut. Work the marinade around the ribs to make sure it's evenly distributed. Place the bags in a deep platter and leave it in the refrigerator for 24 to 48 hours. Turn the meat a few times during this time.

Preheat the oven to 212°F (100°C). Remove the spareribs from the plastic bags; wipe the spareribs clean of the herbs, and season them with salt and pepper. Cut the carrot into large chunks, and the onions in half. Place the vegetables in an ovenproof dish, and set the meat on top of them.

Pour the marinade over the meat and vegetables, place a tight-fitting lid on the dish, and let it cook in the lower half of the oven for about 8 hours. Baste the meat with the pan's juices now and then. Take the meat out of the oven and check for tenderness—if it easily separates from the bone. If so, bump the oven temperature up to 440°F (225°C), baste thoroughly with the pan's juices, and place the meat on the middle rack for approximately 15 minutes to allow the ribs to brown nicely.

Serve the ribs with oven-baked cherry tomatoes. Prick a hole in the tomatoes with a toothpick and brush them with some olive oil. Sprinkle them with some salt flakes and bake them next to the ribs for 15 minutes. Marinated feta (recipe on p. 130) and eggplant salad (recipe on p. 132) make great accompaniments, too.

Brined Shoulder Hock with Cauliflower Mash and Horseradish

Cooked shoulder hock is a true classic from way back; it has been enjoyed in the Nordic countries for a long time. It's a very simple dish to prepare—the trick is to let the meat cook until it is so tender it almost falls apart.

The traditional potato/rutabaga hash has been replaced here by a cauliflower purée with big flavor and fewer carbohydrates.

- 2.2–3.3 lbs. (1–1 ½ kg) brined shoulder hock, bone-in
- water to cover the meat
- 1 large carrot
- 1 yellow onion
- 1 4-inch (10-cm) leek
- 10 white peppercorns
- 6 allspice berries
- 2 bay leaves
- 2.2 lbs. (1 kg) cauliflower
- 2–3 tablespoons horseradish
- 1 ¾ oz. (50 g) butter
- 3 ⅓ fl. oz. (100 ml) bouillon from cooking the meat

Rinse the shoulder hock in cold water. Place it in a stew pot. Cover the hock with water and bring to a boil. Skim thoroughly. Rinse, peel, and cut the vegetables into large chunks, and position them around the meat along with the peppercorns and the bay leaves. Cook covered and on low heat, for about 3 hours, until the meat starts to come away from the bone.

Then, place the cauliflower (separated into florets) in the stock next to the meat and let it simmer for another 10 to 15 minutes. The cauliflower should be soft. Remove the cauliflower with a slotted spoon; place it in a saucepan and mash it with a potato masher. Season with freshly grated horseradish to taste; stir in butter and some bouillon from the meat. Taste and adjust for salt or pepper. Reheat carefully while stirring.

Place the cauliflower mash in deep plates and set slices of meat on top of the mash. Pour some bouillon over the mash and meat, and garnish with chopped parsley.

Cabbage Soup with Pork Shoulder, Fresh Lamb Sausage, and Salsiccia

A very tasty soup made with two different kinds of cabbage, pork shoulder, and fresh lamb and pork sausages is one of the best dishes to serve at dinner on a gray and overcast evening.

I make my soup with homemade chicken stock. We really have a lot of nourishment here in one pot.

- 14 oz. (400 g) pointy, "Hispi," or "sweetheart" cabbage
- 7 oz. (200 g) white cabbage
- 7 oz. (200 g) fresh pork shoulder
- 2 yellow onions
- 1 ¾ oz. (50 g) butter
- 1–2 teaspoons salt
- about ¼ teaspoon white pepper
- 2 teaspoons dried thyme
- 1 teaspoon crushed caraway seeds
- 6 ¾ fl. oz. chicken stock (see p.106) + 2 ½ cups (600 ml), OR 2 cubes organic chicken stock + 3 ⅓ cups water
- 1–2 tablespoons cider vinegar
- 10 Brussels sprouts
- 7 oz. (200 g) fresh lamb sausage
- 7 oz. (200 g) fresh salciccia (a fennel- and anise-infused Italian sausage)

Cut off the thick stem at the bottom of the cabbage and discard; cut the cabbage into slices and then into thick strips. Cut the pork into cubes measuring approximately ½ inch (1 cm) cubed. Peel and slice the onion. In a large, deep pan, fry the diced pork in butter, and season with salt and pepper. Start with the lower amount of salt if you're using stock cubes.

When the pork has browned a little, add the onion and let it fry a bit more. Now add in the white cabbage, the pointy cabbage, thyme, and caraway seeds. Cook, stirring occasionally, for 5 minutes without letting it brown too much.

Add in the stock and vinegar, bring to a boil and let it simmer over low heat for 45 minutes if cooking on the stovetop. In a Crock-Pot set on low, it will take between 4 and 5 hours for the cabbage to become really soft.

Cut the Brussels sprouts into quarters, and add them to the soup along with the sausages. Let the soup simmer for another 15 to 20 minutes on the stovetop, or another hour in the Crock-Pot set on high. Taste and adjust for more tang or more salt.

You can either serve the sausages whole—and set the table with a knife and fork next to the spoon—or cut them into chunks. Serve the soup with a dollop of mayonnaise mixed with some grated horseradish.

LAMB

Turkish Lamb Stew

In Turkish restaurants this dish is often called moussaka. My version here is more like a soupy stew, however, and nothing at all like the Greek meat gratin of the same name that contains potatoes and a cheese sauce made with flour. That's why this is a good dish to include in any low-carbohydrate regimen.

- 1 large eggplant
- 1 lb. 5 oz. (600 g) ground lamb
- 1 ¾ oz. (50 g) butter + 1–2 tablespoons olive oil
- 1 tablespoon dried mint
- 1 tablespoon dried oregano (Mediterranean)
- 1 tablespoon paprika
- about 1 teaspoon salt
- about ½ teaspoon black pepper
- 1 red chili
- 1 red onion
- 3 cloves garlic
- 15 green olives
- 6 ¾ fl. oz. (200 ml) chicken stock (see p. 106) OR 6 ¾ fl. oz. water + ½ organic chicken stock cube
- ⅓ cup + 1 ½ tablespoons (100 ml) Ajvar (a smoked bell pepper relish)

Preheat the oven to 480°F (250°C).

Cut the eggplant in half and set it (peel side up) on a rack high in the oven, and grill it until the halves become charred and turn black; this will take approximately 20 minutes. Turn and grill the other side for 10 minutes. Let the eggplant halves cool, peel them, and cut them into smaller chunks.

Sauté the ground lamb in a stew pot, in half the butter and oil. Season the meat with mint, oregano, paprika, salt, and pepper. Let the spices bloom by frying them with the meat for a little while.

Finely chop the chili, and remove the seeds if you'd rather not end up with a very spicy stew. Peel and chop the onion and garlic, and brown them lightly. Remove the pits from the olives, and cut the olives into pieces; add those pieces to the onion and garlic, and sauté for a few minutes, stirring occasionally.

Mix the onion with the ground lamb, and pour in the chicken stock, eggplant, and the ajvar. Mix well. Bring to a boil, then turn down the heat to the lowest setting and let simmer, covered, for about 1 ½ to 2 hours. Taste while the stew is cooking to adjust for salt. Olives can be quite salty, so season carefully to begin with. Stir occasionally to stop the stew from scorching, and dilute the stew with some water if it gets too thick. This stew is wonderful when served with crumbled sheep's cheese and a dollop of minty haydari yogurt sauce (see recipe on p. 126).

Lamb and Cabbage Stew with Mint, Lemon, and Chili

Lamb and cabbage stew has long been a traditional meal in the Nordic countries. Here I've created a crossover by mixing it with flavors from the Mediterranean kitchen. The mint, lemon, and chili bring this stew a notch above the mundane.

- 8 ½ cups (2000 ml) white cabbage, coarsely chopped
- 4 shallots
- 2 cloves garlic
- 1 organic lemon
- 2.2 lbs. (1 kg) lamb, bone-in
- about 1 tablespoon salt + 1 teaspoon for the cabbage
- about 1 teaspoon red chili flakes
- 3 ½ oz. (100 g) butter + 1–2 tablespoons olive oil
- 1 tablespoon dried mint
- 6 ¾–13 ½ fl. oz. (200–400 ml) water

Cut the white cabbage in half and remove the tough middle stem. Chop the cabbage into pieces 1 ½ inches (4 cm) square. Slice the shallots into thin wedges and the garlic into slices. Wash the lemon and grate the peel finely. Juice the lemon. Brown the meat in some of the butter and olive oil, frying it in batches to avoid crowding the pan, and transfer the chunks to a stew pot. While sautéing, season the meat with salt and chili flakes. Deglaze the frying pan with 3 ⅓ fl. oz. (100 ml) water between the fried batches and pour the pan juices over the meat.

Sauté the white cabbage, onion, and garlic in the butter and olive oil without letting them brown too much; season with salt, mint, lemon peel, and juice of the lemon. Add this to the meat in the pot, and mix thoroughly. Use the larger amount of water if cooking on the stovetop or in the oven; the smaller amount is enough for an electric Crock-Pot.

Cook the meat for about 2 to 3 hours over low heat on the stovetop, or at 300°F (150°C) in the lower part of the oven.

It will take about 6 to 8 hours in a slow cooker set to low, depending on what kind of Crock-Pot you have. Avoid lifting the lid during the first 3 hours to preserve all liquid in the pot. Test to see if the meat is tender—it will readily come away from the bones. Serve this dish in deep plates with aioli or garlic butter. Sprinkle some freshly chopped mint on top for a nice touch.

Lamb Kleftiko

Clay cooking pots were very popular in the 1970s, and they're currently experiencing a well-deserved renaissance. During the Stone Age, clay pots were dug down into cooking holes and surrounded by hot stones. Clay pot cooking is about as near to original cooking methods as we can get today.

Kleftiko, which is made of lamb cooked until it falls off the bone, is a wonderful taste sensation not to be missed while on a holiday in Greece, where it is served with potatoes that have been cooked in the gravy.

I have used eggplant and zucchini instead of potatoes, as they contain very few carbohydrates.

Serves 5 to 6

- 2 eggplants
- 1 zucchini
- 1 red onion
- about ½ tablespoon salt + 1 teaspoon for the vegetables
- about ½ teaspoon black pepper + ¼ teaspoon for the vegetables
- 2 lbs. 10 oz. (1200 g) lamb, bone-in
- 1 tablespoon dried oregano (Mediterranean)
- 1 tablespoon dried thyme
- ½ tablespoon crushed coriander seeds
- 3 ½ oz. (100 g) feta cheese
- 4–6 cloves garlic
- 6 ¾ fl. oz. (200 ml) dry white wine
- juice of one lemon
- 1 ¾ fl. oz. (50 ml) cold-pressed olive oil

Soak the clay pot for at least 15 minutes in cold water, which should cover the pot completely.

Cut the eggplant in half and zucchini lengthwise, and then cut them into ¾-inch (2-cm) slices. Cut the onion and garlic into wedges. Spread the vegetables in the clay pot and season with 1 teaspoon salt and ¼ teaspoon pepper. Turn the vegetables by hand to make sure they're salted on both sides.

Mix oregano, thyme, coriander seeds, salt, and pepper in a bowl. Rub the pieces of meat with the spice mix. Set the meat on top of the vegetables, and drizzle the wine, lemon juice, and olive oil on top. Dice the feta cheese and sprinkle it over the meat.

Cover the clay pot with the lid, and place it in the lower part of a COLD oven (or else the pot might crack). Heat the oven to 350°F (175°C). Let the clay pot cook for 2 to 2 ½ hours, then remove it from oven and test the meat for tenderness. The meat should easily come apart from the bone. BE CAREFUL when you remove the lid from the pot—THE STEAM INSIDE IS EXTREMELY HOT.

Serve the meat with the vegetables, a dollop of crème fraîche or herb aioli (see recipe on p. 122), and a salad of arugula, walnuts, and pomegranate seeds.

Herb-Stuffed Lamb Roulade

Breast of lamb is the lower part of the ribs; it's also called thin brisket or flank of lamb. With a tasty stuffing and slowly roasted in the oven, the end result is just sensational.

Served cold, the lamb roulade is perfect with a glass of wine or a cup of tea in the evening. Simply cut the meat into thin slices and enjoy.

Serves about 6

- 2 tablespoons butter
- ⅓ cup + 1 ½ tablespoons (100 ml) hazelnuts (filberts)
- 7 oz. (200 g) ground lamb
- 3 ½ oz. (100 g) grated Västerbotten cheese (or Parmesan)
- 2 cloves garlic
- 2 large eggs
- 2 tablespoons Dijon mustard
- 2 tablespoons chopped fresh thyme
- 2 tablespoons chopped fresh parsley
- 2 tablespoons chopped fresh chives
- 1.3–1.5 lbs. (600–700 g) boneless breast of lamb
- 1 tablespoon paprika
- about 2 tablespoons salt + 1 teaspoon for the stuffing
- about 2 teaspoons white pepper + ¼ teaspoon for the stuffing
- 3 tablespoons melted butter, for brushing

Preheat the oven to 260°F (125°C). Butter a piece of aluminum foil down the middle with 2 tablespoons butter. Crush the nuts to make coarse flour with a mortar and pestle, or chop finely. Mix the ground meat well with cheese, finely grated garlic, hazelnuts, egg, mustard, chopped herbs, 1 teaspoon salt, and ¼ teaspoon pepper.

Rinse the breast of lamb in cold water, and dry it thoroughly with paper towels. Set it on a cutting board, meat side up. Mix salt, pepper, and paprika, and season the meat with half of this mixture. Spread the stuffing in an even layer over the breast meat.

Roll up the breast of lamb tightly, starting from the thickest short side. Bind up the roll with kitchen twine (see instructions on p. 16).

Season the roulade on all sides and set it on the prepared aluminum foil. Envelop the meat completely with the foil, and fasten the ends by twisting them like the ends of a piece of candy. Place the roulade on a rack in a baking pan, and roast it in the bottom part of the oven for 5 hours. Increase the oven temperature during the last 20 minutes of cooking time to 400°F (200°C), and remove the roulade from the oven. Open up the aluminum foil, but be careful—the steam is hot! Place the meat roll back on the rack, and brush the top of it with melted butter.

Place the roulade back in the oven and cook it until it has browned nicely.

Let the roulade rest under a piece of foil, on the cutting board, for approximately 15 to 20 minutes before cutting it into slices. Serve it with one of the gratins on pp. 144–147, and with one of the herb butters you'll find on pp. 148–150.

Place the roulade in a plastic bag once it has cooled down, and store it in the refrigerator.

GAME

Bacon-Wrapped Caribou Meatloaf

We should consume more caribou meat and other game, as their nutritional profile is far superior to that of grain-fed animals.

Game is lean, so I mix it with ground pork to increase its juiciness and fat content.

- 1 medium yellow onion
- 6 ¾ fl. oz. (200 ml) trumpet (also called winter or funnel) chanterelles, parboiled
- ½ tablespoon crushed juniper berries
- about 1 teaspoon salt
- about ½ teaspoon white pepper
- 1 ¼ lbs. (500 g) ground caribou or other ground game meat
- 3 ½ oz. (100 g) ground pork
- 3 egg yolks
- 1 tablespoon gluten-free tamari soy sauce
- ¼ teaspoon ground ginger
- ¼ teaspoon ground cloves
- ¼ teaspoon ground allspice
- 1 oz. (25 g) butter for frying
- 7 oz. (200 g) bacon

Preheat the oven to 400°F (200°C). Grease an ovenproof dish or casserole.

Chop the onion finely, and the chanterelles coarsely; sauté for a few minutes in butter, and season with juniper berries, salt, and pepper. Let cool. In a bowl, break the ground meat into smaller pieces and add in the egg yolks, soy sauce, and spices.

Add the chanterelles and onions, and mix it all quickly together. Test-fry a small piece and taste for seasonings. Keep in mind that the bacon will add a good bit of salt to the meatloaf.

On a cutting board, place half the bacon slices side by side each to make one long square. Using the remaining bacon, weave one slice at a time straight across by lifting every second slice over the new slice.

Wet your hands with cold water and shape the meat mixture into a chubby, rectangular meatloaf. Set the loaf on top of the bacon, and drape the slices across the loaf so they lie somewhat crossed on top. Press lightly with your hands to make sure that the bacon stays put around the loaf. Turn it carefully and place it in the prepared oven dish or casserole; cover with a lid or aluminum foil and set the loaf in the lower part of the oven. Bake covered for about 30 minutes. Remove the lid or foil to allow the loaf to develop a nice crust in the last 10 to 15 minutes of cooking.

Brussels sprouts fried with mushrooms and red onion, as well as herb butter, are nice sides for this dish. The recipe for the butter is on p. 148.

The meatloaf is very tasty even when cold, so it makes an excellent substitute for a sandwich. If using it as such, slice the loaf thinly and serve it with butter and cheese.

৵৩

Swedish Venison Tjälknöl

Swedish tjälknöl, or "frost lump," is named thusly because the meat starts out frozen hard. There is no better meat than game, especially for all of us who follow the natural, low-carbohydrate nutritional guidelines featured in LCHF/Paleo diets. Animals destined for meat production who are left to graze in the fields—feeding on what they are biologically meant to eat—are far less stressed than their counterparts in the factory farm, which is evident from the taste of their meat.

In Sweden, game can often be found in well-stocked grocery stores and markets; that is, if you are not fortunate to have relatives who are hunters.

- 2.2 lbs (1 kg) frozen roast of venison, deer, caribou, or elk
- 1 large carrot
- 2 tablespoons juniper berries
- 1 teaspoon whole white peppercorns
- 4 ¼ cups (1000 ml) water
- ⅓ cup + 1 ½ tablespoons (100 ml) salt
- 3 star anise
- 1 bay leaf, crumbled

Place the frozen roast on a rack in an ovenproof dish. Set the dish in the lower part of the oven, and heat the oven to 212°F (100°C). After about 3 hours, when the meat has defrosted a bit, stick a meat thermometer into its thickest part.

Roast the meat for approximately 8 to 10 hours, depending on the thickness of the roast. When the thermometer shows 149°F (65°C) the meat will be pink. If you prefer it more well done, let the temperature reach 158°F (70°C).

Peel and coarsely grate the carrot. We're using carrot to add some sweetness to the stock instead of the sugar that is typically added when seasoning a "frost lump." Crush the juniper berries and white peppercorns with mortar and pestle. Mix all ingredients for the stock in a saucepan and bring it to a boil. Place the meat in a high-sided, narrow bowl or container; pour the hot stock over the meat. Make sure it covers all of the meat. Let the meat cool and leave it in the refrigerator for about 4 to 5 hours.

Remove the meat from the liquid, let it drain in a colander, and dry it with paper towels. Cut the roast into thin slices

Serve the cold meat with fennel coleslaw (recipe on p. 138) or one of the gratins (pp. 144–147).

Boar Roast Braised with Apples

When boar meat is braised in apple cider with apples at low temperature, it develops a very full and robust flavor. Wild pig is stronger-tasting than its domesticated counterpart; the meat is especially well paired with the sweetness of apples.

Adding a splash of Calvados (apple brandy) in the gravy accentuates the delicious taste of apple yet further.

Serves 5 to 6

- 1 boar roast, about 2.2 lbs (1 kg)
- about 1 tablespoon salt
- about 1 teaspoon white pepper
- 1 carrot
- 1 4-inch (10-cm) leek
- 2 apples
- 1 ¾ oz. (50 g) butter
- 1 ¾ fl. oz. (50 ml) veal stock (see p. 170) + 3 ⅓ fl. oz. (100 ml) apple cider, OR ½ cube organic meat stock + 5 fl. oz. (150 ml) apple cider
- 1 ¾ fl. oz. (50 ml) Calvados (apple brandy)

If the boar roast is not already bound in a net, bind it with kitchen twine—you'll find instructions for how to do this on p. 16. Rub the meat with salt and pepper.

Brown the roast well on all sides. Peel and slice the carrot and cut the leek into chunks. Quarter the apples, leaving their peels on.

Sauté the vegetables with the meat for a little bit. Pour in the veal stock, cider, and Calvados. Bring to a boil, and let it braise (i.e., let it cook lightly in the liquid), covered, for 2 to 2 ½ hours. Turn the meat over a few times while it's cooking so the flavorful gravy bastes the entire roast. Test the meat for doneness with a carving fork—the fork should slide in easily.

Serve thick slices of the roast with sauerkraut warmed with some crispy bacon, juniper berries, and leek. The recipe for the mustard butter that goes with this dish can be found on p. 150.

Or, make a smooth cream sauce with the richly flavored gravy. Reduce the pan's juices by cooking them down to 6 ¾ fl. oz. (200 ml) and add 2 cups (500 ml) heavy cream. Let simmer for at least 45 minutes, until the sauce has thickened. Season with salt, pepper, and some more Calvados.

Elk Stew with Forest Mushrooms

A stew full of enticing aromas from the woodland's bounty is very welcome on a raw and chilly evening. Elk meat has a full flavor that blends well with smoked pork belly, and if you have the opportunity to collect the mushrooms yourself, it'll add that much more to your enjoyment of the meal. I have used chanterelles and trumpet (also called winter or funnel) mushrooms that I collected on my trips in the fjeld forests in the north of Sweden.

- 1 teaspoon crushed juniper berries
- 1 teaspoon dried thyme
- 1 lb. 5 oz. (600 g) elk meat, or other game, cut into 1 ¼-inch (3-cm) cubes
- ½–1 tablespoon salt
- about 1 teaspoon black pepper, coarsely crushed
- 1 ¼ cups (300 ml) forest mushrooms, parboiled
- 1 ¼ cups (300 ml) leek, cut into coarse strips
- 1 ¾ oz. (50 g) hot-smoked pork belly, cut into thin strips
- 3 ½ oz. butter, for frying
- 3 ⅓ fl. oz. (100 ml) veal stock (recipe see p. 106) + 3 ⅓ fl. oz. (100 ml) water OR ½ organic meat stock cube + 6 ¾ fl. oz. (200 ml) water
- 1 tablespoon Dijon mustard
- 1 ¼ cups (300 ml) heavy cream

Work the juniper berries and dried thyme into a coarse powder with a mortar and pestle. Rinse the meat in cold water and let it drain. Dry the chunks with paper towels, and brown them by frying them in several batches. Season them with salt, pepper, juniper berries, and thyme. Use the smaller amount of salt to begin with, as the smoked pork belly will also be salty.

Place the meat in a stew pot. Deglaze the frying pan with some water between frying the batches, and pour the juice over the meat. Clean the frying pan with a paper towel.

Now brown the mushrooms and leek, and add them to the meat. Finally, fry the strips of pork belly. Stir everything into the stew pot, and mix well with stock and mustard. Bring to a boil, and turn down the heat and simmer, covered, for 2 hours. Stir occasionally.

Add the heavy cream and bring back to a boil. Let the contents boil, uncovered, until the sauce thickens and the meat is very tender—this usually takes about half an hour. Taste and adjust for salt or pepper. Serve the stew with cooked broccoli and cranberries.

POULTRY

Whole Roasted Chicken

A large chicken that has slowly absorbed the flavors of lemon, chili, garlic, and cilantro will outshine any glutamate-enriched rotisserie chicken any day. Here I've cooked the bird in a clay pot, which makes it even more juicy and delicious.

- 1 organic chicken, 2.65–3.3 lbs. (1200–1500 g)
- 2 organic lemons
- ½ package fresh cilantro
- 2 cloves garlic
- about 1 teaspoon red chili flakes
- about 1 tablespoon salt
- 1 ¾ oz. (50 g) butter, for brushing
- about 1 tablespoon paprika
- kitchen twine
- 1 ¼ cups (300 ml) heavy cream or coconut cream

Soak the clay pot for at least 15 minutes in enough cold water to cover the pot. Rinse the chicken thoroughly, inside and out, and let it drain. Wash the lemons and cut them into eighths. Snip off a bunch of cilantro and chop it coarsely. Mix the cilantro and lemon in a bowl. Peel and quarter the garlic cloves. Mix the garlic and chili flakes with the lemons.

Salt inside the cavity of the bird, and then stuff it with the lemon mixture. Brush the exterior of the chicken with melted butter. Season it with the remaining salt and paprika. Bind the legs and wings with kitchen twine, and place the chicken in a greased clay pot.

Cover the pot with a lid and place the pot in a COLD oven. Turn the oven up to 350°F (175°C). Roast in the chicken in lower part of the oven about 1 ½ to 2 hours.

If you don't have a clay pot, you can roast the chicken in the oven like this:

Set the chicken on a rack in small ovenproof baking dish. Roast it in the lower part of the oven for 2 ½ to 3 hours at 300°F (150°C). Brush the chicken with melted butter a few times while you roast it. Toward the end of cooking, pour in 6 ¾ fl. oz. (200 ml) of water to loosen the pan's juices.

After 2 ½ hours of roasting time, insert a wooden toothpick in the chicken to check if the bird is ready. There should be no pink meat juice—it's unlikely that there would be any after such a long time in the oven. To make things easier, use a meat thermometer; it should read 176°F (80°C) when the chicken is done.

Remove the chicken, cover it loosely with aluminum foil, and let it rest for a while before carving. Pour the pan juices into a saucepan and bring to a boil. Add in 1 ¼ cups (300 ml) heavy cream or coconut cream, and boil until you have a smooth, creamy sauce. Season it with salt, chili, and some lemon juice. Serve the chicken with the sauce, zucchini, or slices of eggplant sautéed in butter, and a salad of radishes and goat cheese (see recipe on p. 136).

Butternut Squash Soup with Chicken, Coconut, and Ginger

Not only does it display a vibrantly beautiful color, butternut squash also contains a lot of beneficial nutrients, which is why this vegetable should become a regular feature on our tables. I often buy butternut squash when I'm in Turkey and roast it in the oven with herbs and butter.

I've made the recipe completely dairy-free. It's a truly wholesome soup made from coconut milk and flavored with ginger, garlic, and turmeric.

- 1 butternut squash, about 2.2 lbs. (1 kg)
- 3 ⅓ fl. oz. (100 ml) coconut oil (with coconut flavor)
- ½–1 teaspoon salt
- about ½ teaspoon red chili flakes
- 2-inch (5-cm) piece of fresh ginger
- 2 cloves garlic
- 6 ¾ fl. oz. (200 ml) finely chopped leek, white only
- 1 tablespoon turmeric
- 3 ⅓ fl. oz. (100 ml) chicken stock (see p. 106) OR 1 organic chicken stock cube + 3 ⅓ fl. oz. (100 ml) water
- 2 cups (500 ml) coconut cream
- 13 ½ fl. oz. (400 ml) water
- juice from 1 lime
- 1 ⅔ cups (400 ml) shredded cooked chicken

Preheat the oven to 350°F (175°C).

Cut the squash into wedges, scrape out the seeds, and cut scoop out the flesh from the peel. Cut the flesh into smaller chunks and place them into an ovenproof dish that has been brushed with coconut oil. Toss the chunks of squash with coconut oil and season with salt and chili flakes. Start with the smaller amount of salt, as the stock also contains a bit of salt. Roast the squash for 30 minutes.

Peel the ginger and garlic; grate them finely. Heat the rest of the coconut oil in a stew pot, and sauté the ginger, garlic, and leek for a few minutes without letting them brown. Add the turmeric and sauté a bit longer. Add in the now-soft chunks of squash, and pour the chicken stock or stock cube and water over the contents of the pot. Stir in the coconut cream and bring to a boil.

Leave the soup to simmer for half an hour. Blend it smooth directly in the saucepan by using an immersion blender, or use a food processor. Season the soup with some lime juice, and more salt and chili if needed.

Heat the shredded chicken in a covered saucepan with a small amount of stock. Place the meat in four bowls and pour the steaming hot soup over it. Garnish the bowls with red chili flakes and some fresh cilantro. Serve the soup with a dairy-free muffin (see recipe on p. 118).

Indian Korma

While most of the dishes prepared in my kitchen feature Nordic or Mediterranean flavors, Indian comes in a close third among my favorite cuisines of the world. Korma is a mild stew that contains coconut, cream, and almonds. There is often sugar and raisins, too, but they are no-shows when you eat according to LCHF principles. I promise you that the korma is still utterly delicious.

Here I have made a completely dairy-free stew containing only coconut cream. However, you can use half coconut cream and half heavy cream if you prefer a milder coconut flavor.

- 1 lb. 5 oz. (600 g) turkey breast
- 3 ⅓ fl. oz. (100 ml) coconut oil (intense coconut flavor)
- about ½ tablespoon salt
- about ¼ teaspoon cayenne pepper
- ⅓ cup + 1 ½ tablespoons (100 ml) fresh ginger, sliced
- 1 yellow onion
- 3 cloves garlic
- 1 tablespoon green cardamom, whole pods
- 1 tablespoon ground coriander
- 1 teaspoon turmeric
- 1 teaspoon cumin
- 1 teaspoon cinnamon
- ⅓ cup + 1 ½ tablespoons (100 ml) almond flour
- 2 cups (500 ml) coconut cream
- juice of ½ lemon

Cut the turkey meat into 1 ¼-inch (3-cm) cubes. Sauté the pieces in half the coconut oil until lightly browned. Season with salt and cayenne, and set aside.

Coarsely chop and mix the ginger, onion, and garlic together. If you don't have a food processor, chop finely by hand.

Sauté these seasonings for five minutes in the remaining coconut oil in a stew pot over medium heat, stirring frequently to prevent it from burning. Crush the cardamom pods lightly in a mortar and pestle, but leave some pods whole. Sprinkle them over the onion mix together with the other spices and the almond flour. Let this mixture—now dry—cook for a few minutes on the lowest setting. Stir from bottom of the pan continuously to prevent it from scorching.

Add in the turkey and coconut cream. Bring to a boil, and let simmer over low heat for about 1 ½ hours to allow the meat to become very tender and give the sauce enough time to thicken. If the sauce is too thick, thin it with some water. Season with lemon juice and some more salt, if needed. Serve with flaked, roasted coconut, and a salad of tender chard leaves with some pomegranate seeds.

Chicken Pie in an Almond Flour Crust

I have tried many different ways to develop a worthy gluten-free piecrust for savory pies.

This particular version is made with almond flour, and it is the best of the bunch so far. It has a sturdy bottom, it's very tasty and, of course, it has many more good-for-you ingredients than a traditional wheat flour crust.

Almond Flour Piecrust

- 1 ¼ cups (300 ml) almond flour
- 1 tablespoon psyllium husk powder
- ¼ teaspoon salt
- 1 ¾ oz. (50 g) butter, chilled
- 1 large egg

Mix almond flour, psyllium husk powder, and salt in a bowl. Cut the butter into small cubes, place them in the flour, and add the egg.

Work the dough together quickly until the butter is well blended with the flour. The dough will be quite loose, so move it into the pie pan and spread it with the help of a rubber spatula.

Place a piece of cling film over the pie pan. To get an even crust, use your fingers to press the dough over the bottom of the pan and up around the edges.

Place the pie pan in the refrigerator to firm up for 30 minutes. Then remove the cling film and prebake the crust at 400°F (200°C) for 5 minutes.

Chicken Pie

- 1 ¾ oz. (50 g) butter
- 7 oz. (200 g) button mushrooms, sliced
- 1 4-inch (10-cm) or 6 ¾ fl. oz. (200 ml) sliced leek
- about 1 teaspoon salt
- about ½ teaspoon red chili flakes
- 1 teaspoon dried French tarragon
- 1 teaspoon dried chervil
- 1 teaspoon dried parsley
- 1 tablespoon white balsamic vinegar
- 1 tablespoon gluten-free soy sauce
- 14 oz. (400 g) cooked chicken, chopped

Brown the butter slightly; sauté mushrooms and leek in the butter for a few minutes so they brown a bit. Season them with salt, chili, and herbs. Add in the soy sauce and vinegar, and cook to reduce. Add in the chopped chicken, and sauté for a few minutes.

Place the filling in the prebaked piecrust, and pour in the savory egg custard (recipe follows).

Savory Egg Custard

- 3 large eggs
- 6 ¾ fl. oz. (200 ml) crème fraîche
- ⅓ cup + 1 ½ tablespoons (100 ml) grated aged cheese
- about ½ teaspoon salt
- about ¼ teaspoon white pepper

Whisk the eggs with the crème fraîche and the cheese; season with salt and pepper. Pour the egg custard over the filling in the crust, and spread it evenly over the pie.

Bake for 30 to 40 minutes at 350°F (175°C).

French Tarragon Chicken with Smoked Pork Belly and Olives

Chicken, mushrooms and French tarragon are flavors that go very well together. Black olives and smoked pork belly impart a lovely salty and slightly smoky note to the chicken.

Now we're talking real slow cooking, as this stew will simmer for a whole day in the electric Crock-Pot. If you haven't got a slow cooker, however, the stew is just as delectable if prepared slowly on the stovetop or in the oven.

- 3 bell peppers in a variety of colors
- 1 ¼ cups (300 ml) coarsely julienned leek
- 7 oz. (200 g) small portabella mushrooms
- 3 ½ oz. (100 g) hot-smoked pork belly
- 1 ¾ oz. (50 g) butter
- about ½ tablespoon salt
- about 1 teaspoon coarsely ground black pepper
- 2 tablespoons dried French tarragon
- 1 tablespoon gluten-free tamari soy sauce
- 5 fl. oz. (150 ml) dry white wine
- 3 ⅓ fl. oz. (100 ml) chicken stock (see recipe p. 106) OR 1 organic chicken stock cube + 3 ⅓ fl. oz. (100 ml) water
- 4 chicken legs
- 6 ¾ fl. oz. (200 ml) black olives, with pits

Preheat the Crock-Pot insert.

Cut the bell peppers and leek into strips. Quarter the mushrooms. Slice the pork belly and cut the slices into strips. Brown the mushrooms in butter; season lightly with salt and pepper. Add in the bell peppers and the leek, and sauté for a few minutes. Crumble in the tarragon; add the soy sauce, wine and stock. Bring to a boil.

Place an even layer of vegetables over the bottom of the Crock-Pot. Sauté the chicken legs for a few minutes on both sides; season with salt and pepper. Place them on top of the vegetables in the Crock-Pot.

Brown the pork belly, and scatter the pieces, along with the olives, all over the chicken. Place the lid on the Crock-Pot and turn the setting to low. Leave the stew to simmer for 6 to 8 hours, depending on the thickness of the chicken legs. To check for doneness, stick them with a sharp knife in the thickest part of the leg to see if the meat is tender, and to ensure that any running juice is not pink.

This stew can also be cooked on the stovetop over 2 to 3 hours on the lowest heat setting or flame. If cooked in a casserole with a lid, in an oven set at 260°F (125°C), it will take 3 to 4 hours.

Serve with aioli or garlic butter, and with a green salad.

FISH AND SHELLFISH

Seafood Stew

My tasty fish stew is fit both for an everyday meal and for a more festive occasion. It's best to eat fish and shellfish often, as they're loaded with beneficial omega-3s.

Makes approximately 6 servings

- 2.2 lbs. (1 kg) mussels
- 6 ¾ fl. oz. (200 ml) water
- 3 ⅓ fl. oz. (100 ml) dry white wine
- ⅓ cup + 1 ½ tablespoons (100 ml) julienned leek
- 10 whole white peppercorns
- ½ teaspoon salt

Wash the mussels in running water. Throw away any damaged ones, as well as those that don't close when you press the shells together. Bring the water and wine, together with the leek, peppercorns, and salt, to a boil. Add the mussels, cover the pot with a lid, and boil rapidly at high heat for 5 minutes.

Remove the pot from the stovetop and discard any unopened mussels.

- 1 ¼ lb. (500 g) salmon fillet
- 1 ¼ lb. (500 g) pollock
- juice of 1 lemon
- about 1–2 teaspoons salt
- 2 beefsteak tomatoes
- 2 tablespoons fennel seeds
- 2 tablespoons anise seeds
- 1 packet saffron threads
- about 1 teaspoon chili flakes
- 2 fennel bulbs
- 2 carrots
- 1 yellow onion
- 1 ¼ cups (300 ml) julienned leek
- 1 ¾ oz. (50 g) butter, for frying
- 3 ⅓ fish stock (see p. 108) + 6 ¾ fl. oz. (200 ml) dry white wine OR 1 organic fish stock cube + 6 ¾ fl. oz. dry white wine + 3 ⅓ fl. oz. (100 ml) water
- 7 oz. (200 g) shrimp in the shell

Cut the fish into smaller pieces, and squeeze lemon juice over them. Salt lightly and mix well.

Scald the tomatoes for a minute in boiling water. Remove the tomatoes from the water, peel off their skins, and cut the flesh into smaller pieces.

Crush the fennel seeds, anise seeds, saffron threads, and chili flakes with a mortar and pestle. Slice the fennel bulbs and carrots thinly, and chop up the onion. Fry all the vegetables, seasoning with salt and pepper.

Transfer the vegetables to a Crock-Pot or to a stew pot. Add in the fish stock, wine, and spices. Let cook for 4 to 5 hours in the Crock-Pot, set to low. In a stew pot on the stovetop, it will take about 1 ½ hour to cook, over low heat. Carefully add the fish toward the end of the cooking time, and let it simmer, covered. This will take 30 minutes in the Crock-Pot, but only 5 minutes on the stovetop. Add the mussels and shrimp to heat right before serving. Taste and adjust seasoning for more wine or salt.

Serve the seafood stew with aioli (see recipe on p. 122).

Fish and Scallop Paté

This smooth fish paté can be served cold as a delicate appetizer, or warm as an entrée with a white wine or shrimp sauce.

It's also delicious with a filling of shrimp, smoked salmon, or mussels.

Serves 8 as an appetizer

- juice from 1 lime
- 8 scallops
- 1–1 ½ teaspoons salt
- 1 ¼ lbs. (500 g) boneless fish fillet
- 5 ¼ oz. (150 g) butter
- 2 large egg yolks
- 1 ¾ fl. oz. (50 ml) dry white wine
- 1 tablespoon psyllium husk powder
- 1 ¼ cups (300 ml) heavy cream
- 6 ¾ fl. oz. (200 ml) finely chopped chives
- 3 ⅓ fl. oz. (100 ml) finely chopped dill
- about ¼ teaspoon cayenne pepper
- 2 large egg whites

Preheat the oven to 300°F (150°C) and place a rimmed baking sheet, ¾ filled with water, in the lower part of the oven.

Press the lime juice over the scallops and salt lightly. Cut the fish into smaller pieces and process until smooth in a food processor, or grind finely in a meat grinder. Quickly add in the butter, egg yolks, and white wine. Pour into a large bowl.

Stir in psyllium husk powder, cream, herbs, spices, and the lime juice (from the scallops), and mix well with a handheld electric mixer. Beat the egg whites until stiff peaks form, and fold into the paté mixture. Taste and adjust for seasonings.

Cover a rectangular, 1 ½ quart (1500 ml) dish, 11 inches x 7 inches x 1 ½ inches, with plastic wrap. The plastic wrap can handle the oven's heat when it is cooked in a water bath (bain-marie). Fill the dish with paté mixture halfway, and then place the scallops in a row down the middle. Add the remaining paté around and on top of the scallops. Press down on the paté with a rubber spatula. Smooth the surface and wrap with plastic wrap, pressing down a little with the palm of your hand. Cover the dish completely with foil, and set the pan in the water bath.

Bake for 1 ½ hours, or until a thermometer shows 167°F (75°C). Remove the paté from the oven and let cool in the pan. Pour off any condensation that formed while cooling before setting the pan in the refrigerator. Place a weight on top of the paté – this will help it firm up and make it easier to cut into attractive slices. Leave the paté in the refrigerator overnight.

Slice the paté with a thin, sharp knife. Dip the knife in warm water between slices to make even, clean cuts. Leftovers can be frozen, and defrosted slowly in the refrigerator for another occasion. Roe sauce (see recipe on p. 128) pairs very well with this paté.

Apple-Cured Salmon

Swedish cured salmon traditionally contains a lot of sugar. Here, I have cured the salmon using the sweetness of apples instead, which is in fine harmony with the salty flavor.

Always cure salmon that has been frozen for at least 3 days—this ensures that the fish is truly safe to eat, even though it is raw.

- 1 ¼ lbs. (500 g) salmon fillet, middle section, preferably wild caught
- 3 tablespoons salt
- 1 tablespoon white peppercorns, coarsely crushed
- 2 red apples
- 1 bunch dill
- 2–3 tablespoons Calvados (apple brandy), optional

Using tongs or tweezers, remove the rib bones along the thick back section on the salmon. Mix the salt and white pepper. Grate the apples coarsely, and chop the dill in smaller pieces, both sprigs and leaves.

Mix the apples and dill, and place half of it on the bottom of a dish with at least an inch tall rim around it. Rub half of the salt mix into the skin side of the salmon, and set it on top of the dill mixture. Rub the flesh with the rest of the salt, and spread the apple and dill over it evenly.

If you want your salmon to have a more intense apple taste, add some Calvados.

Cover the salmon with plastic wrap. Place a cutting board over it and set a weight on top—a large mortar or a jar of pickles, for example. Leave the salmon to cure for two days, turning it a few times while it cures.

Once the salmon is ready, scrape off all the dill and apple. Cut the salmon on the bias into thin slices with a sharp knife. Serve it with a dollop of mayonnaise (see recipe on p. 124) mixed with Dijon mustard and finely chopped dill, and perhaps a piece of walnut bread (recipe on p. 112).

Fish for a Feast: Gratin with Cod and Shrimp

This is a truly festive dish, and a dream come true for those of us who eat food with few carbohydrates but lots of natural fats. The top layer—made from egg yolks, butter, almond flour, grated cheese, and cream—packs a lot of nutrition, even when used with a leaner type of fish.

Naturally, this dish is excellent when made with salmon, too, in which case it will also be loaded with omega-3 fats.

- 1 lb. 5 oz. (600 g) cod fillet or other fish fillet
- about 2 teaspoons salt
- about 1 teaspoon white pepper
- 3 large egg yolks
- 2 ½ oz. (75 g) butter, softened
- ¼ cup (50 ml) almond flour
- 1 ¼ cups (300 ml) aged cheese, grated (Emmentaler, for example)
- 1 ¼ cups (300 ml) heavy cream
- 7 oz. (200 g) cooked shrimp, peeled

Preheat the oven to 400°F (200°C). Rinse the fish in cold water, and let it drain in a colander. Dry it with paper towels. Season it with salt and pepper on both sides, and set it in a greased ovenproof dish. Mix together the egg yolks with butter, and brush over the fish. Sprinkle the almond flour evenly over the fish fillets. Bake in the middle of the oven for 10 minutes to brown the top layer a little. Remove the fish from the oven and lower the oven heat to 350°F (175°C). Sprinkle the grated cheese over the top of the fish, pour on the heavy cream, and place the dish back in the oven for 30 minutes.

Garnish the fish with peeled shrimp. Cooked cauliflower and a salad of arugula, tomato, asparagus, leek, and lemon-flavored olive oil both make fine side dishes.

Baked Char with Butter-Sautéed Oyster Mushrooms

Whole, wild-caught fish becomes very juicy and flavorful when it's baked in the oven at very low temperature. Here, the fish is served with a delicious mix of mushrooms, bell peppers, and spring onions simmered in butter and lemon.

Mushrooms

- 8 ¾ oz. (250 g) oyster mushrooms
- 1 bunch scallions
- 1 yellow bell pepper
- 1 red chili pepper
- 5 ¼ oz. (150 g) butter
- about 1 teaspoon salt
- juice from 1 lemon

Char

- 2.2 lbs. (1 kg) char or other fatty fish
- about 2 teaspoons salt
- about 1 teaspoon white pepper
- 1 ¾ oz. (50 g) butter
- a few sprigs of dill

Clean and cut the mushrooms into two or three pieces, depending on their size. Cut the scallions in three chunks. Cut the bell pepper in half, and cut it into coarse diagonal slices. Seed and finely julienne the chili pepper.

Preheat the oven to 257°F (125°C). Brown 3 ½ oz. (100 g) butter; sauté the mushrooms until they get a bit of color, and season with half the salt and chili. Transfer this mix to the bottom of an ovenproof dish. Sauté the scallions, bell pepper, and chili lightly in the remaining butter; add salt and lemon juice. Spoon the vegetables over the mushrooms.

Rinse the fish well and dry it with paper towels. Season it inside and out. Cut the butter into pieces and stuff them in the fish along with the dill. Set the fish on top of the mushroom mix, and bake in the middle of the oven for 30 minutes. Remove the dish from the oven and baste the fish with the melted butter a few times during the cooking time.

Bump the oven's heat up to 400°F (200°C) and let the fish brown a little; this will take about another 10 minutes. A thermometer in the thickest part of the fish will show 131°F (55°C) when it's ready.

Serve the char with a green salad and a sauce made with roe (see recipe on p. 128).

ORGAN MEATS

Swedish Blood Pudding

There probably aren't many people today who realize that they can make blood pudding at home in their very own kitchens. I still remember the taste of the blood pudding made from elk blood that my mother used to cook in a water bath. It's a delicious sense memory from my childhood.

Here I have made a gluten-free version of blood pudding that contains all the traditional seasonings, save for the sugar. By using an apple cut into small cubes and coconut flour, a slight sweetness still comes through. The nutritional value of the pudding is also completely different from its store-bought, flour- and carbohydrate-laden counterpart.

Sauerkraut mixed with raw cranberries is both a tasty and nutritious dish. I got this tip from the Swedish low-carbohydrate blog Vid vedspisen (In Front of the Wood Stove).

Serves 8 to 10

- 3 ⅓ fl. oz. (100 ml) veal stock (see p. 106) + 1 ¼ cups (300 ml) water OR 1 organic (beef) stock cube + 1 ⅔ cups (400 ml) water
- 1 red onion
- 1 tart apple
- 5 ¼ oz. (150 g) unrendered fatback lard, diced
- 1–1 ½ tablespoons salt
- ½ teaspoon ground cloves
- ½ teaspoon ground ginger
- ½ teaspoon white pepper
- ⅛ teaspoon ground allspice
- 4 ¼ cups (1000 ml) pig or cow blood
- 1 ¼ cups (300 ml) coconut flour
- 4 tablespoons psyllium husk powder

In Sweden you can buy unrendered lard (often frozen), with or without rind, in well-stocked grocery stores. Slice it thinly and then dice it into fine cubes. Hang the rind outside to make the birds happy!

Preheat the oven to 300°F (150°C) and set on its lower rack a high-rimmed baking pan, filled two-thirds with water.

In a saucepan, bring the stock to boil and then let it cool. Peel and finely dice the onion and apple. Heat the diced lard in a frying pan over medium heat. Once the fat begins to melt, add in the onion and apple together with ½ tablespoon salt and the spices. Fry for about 5 minutes, stirring often and taking care to not let the mixture brown. Let it cool.

Whisk the blood vigorously into the stock, and strain it into a bowl. Mix the coconut flour and psyllium husk powder, and sift it into the blood and stock mix. Blend this mixture thoroughly with a handheld electric mixer. Let the batter rest for 10 minutes. Once the batter has thickened, thoroughly stir in the onion and apple. Fry a dollop of the batter, taste, and adjust for seasoning.

Line two 1-quart (1-liter) loaf pans with a layer of plastic wrap—the wrap can take the oven's heat when it's used in a water bath (bain-marie), and it makes it easier to

unmold the pudding by preventing it from sticking to the loaf pan.

Pour the batter into the loaf pans. Butter or oil two pieces of foil, identical in size to the loaf pans, down the middle, and cover the puddings well. Carefully place the loaf pans in the now-warm water bath, and let them bake for 2 hours.

Remove the pans from the oven, and test the puddings with a toothpick to see if they're firm and dry. Remove the foil, let the pudding cool, cover with plastic wrap, and set it in the refrigerator overnight.

Unmold the blood pudding, remove the plastic wrap, and cut the pudding into ½-inch (1-cm) thick slices. Dip the knife in hot running water between each slice to make the cuts cleaner.

Fry the slices of pudding in butter, and serve with fried pork belly, melted butter, and sauerkraut mixed with raw cranberry preserves or lingonberries.

Freeze the leftover pudding in one block for an easy and ready-to-fix dinner.

Oven-Baked Calf's Liver with Onions and Bacon

Calf's liver baked in the oven at a very low temperature is a true delicacy for those of us who enjoy nutritious pieces of organ meat.

Soak the liver in milk overnight if you prefer milder tasting liver. My mother prepares elk liver this way, and the result is always delicious.

- 1 lb. 5 oz. (600 g) calf's liver
- 3 tablespoons butter
- about 1 teaspoon salt
- about ¼ teaspoon white pepper
- 1 ¾ fl. oz. (50 ml) water

For Flavor
- 3 ½ oz. (100 g) bacon
- 2 red onions, cut into thin wedges
- 1 oz. (25 g) butter, for frying
- a pinch of salt
- about ¼ teaspoon white pepper
- 1 tablespoon apple cider vinegar
- capers
- melted butter
- lingonberries or cranberries

Preheat the oven to 300°F (150°C). Rinse the liver in cold water and remove any veins and outer membrane. Brown the butter in a large stew pot, and fry the liver until it has browned nicely all around. Season it with salt and pepper. Add some water to prevent the liver from drying out in the oven. Cover the liver with a few slices of bacon—it adds great flavor and protects the liver's surface from drying out.

Cover the pot with a lid, and set it in the lower half of the oven. Let it bake until the temperature inside the liver reaches 158°F (70°C). This will take about 1 hour, depending on how thick the liver is, but make sure not to bake the liver for too long or it will become dry and tough.

Place the liver on a cutting board, cover it with foil, and let it rest while you fry the onion wedges and strips of bacon in butter; add the bacon that you used in the oven, too. Season this lightly with salt and pepper (bacon is often very salty). Pour in the vinegar and let the sauce cook down and reduce.

Serve the liver in slices, with the bacon hash and melted butter. Garnish with capers and lingonberries—or cranberries if you can't find lingonberries.

Chicken Liver Paté

This mild paté is delicious served as a starter course, accompanied by a very fresh and tart sea buckthorn jelly.

If you prefer paté with a more classic liver flavor, skip the wine and add 3–4 small anchovy fillets instead.

- 3 ½ oz. (100 g) butter
- 8 ¾ oz. (250 g) chicken livers
- 7 oz. (200 g) chicken breast
- about ½ tablespoon salt
- ½ teaspoon white pepper
- ¼ teaspoon cayenne pepper
- 1 tablespoon psyllium husk powder
- 1 ¾ fl. oz. (50 ml) dry sherry, port wine, or Madeira
- 6 ¾ fl. oz. (200 ml) smetana (or full-fat Greek yogurt)
- 2 large egg whites

Sea Buckthorn Jelly

- 3 ⅓ fl. oz. (100 ml) frozen sea buckthorn berries (or pomegranate seeds)
- 2 sheets of gelatin (or about 1 ½ teaspoons powdered gelatin)

Preheat the oven to 300°F (150°C). Prepare a water bath (bain-marie) for the paté to bake in. Place a deep ovenproof dish filled three-quarters full with warm water in the lower part of the oven.

Melt the butter and let it cool. Rinse the liver and chicken breast in cold water, and let them drain thoroughly in a colander. Once drained, set them on paper towels so all their remaining liquid is absorbed. Cut the liver and chicken in small chunks.

In a food processor, mix the meat together with spices, psyllium husk powder, and wine to make a smooth paste. Quickly add the smetana (or yogurt) and egg whites. If you want a perfectly smooth paté, press it through a strainer.

Fry a silver-dollar-size piece of the paté in some butter, taste it, and adjust for seasoning. Cover a terrine or a small ovenproof dish (about one quart in size) with plastic wrap, and fill it with the paté. Cover the top completely with plastic wrap, and add either some aluminum foil or a lid to cover.

Bake the terrine for 1 ½ hours. If you use a thermometer, it will read 176°F (80°C) when the paté is ready. Set the dish in a sink with cold water to hasten the cooling process, then set the dish in the refrigerator overnight, before it can be unmolded the next day. Cut into several smaller pieces, and freeze whatever is not eaten in the next few days.

Serve up the paté with a spoon dipped in warm water, or slice it with a sharp knife rinsed in warm water.

Sea Buckthorn Jelly:

Soak the sheets of gelatin in cold water for 5 minutes (if using powder, follow the instructions on back of the packet). In a saucepan, warm the berries and stir in the drained gelatin leaves. Pour the mix into a small dish and leave it for a few hours in the refrigerator to set. Remove the jelly with a spatula and cut it into small squares.

STOCKS AND BOUILLONS

Chicken Stock

Making chicken stock is very simple, even if it does take a bit of time. If you cook it from a whole chicken and not just from the carcass of the bird, your dinner will also be ready.

You only need to pick the meat off the bones, which you could then, for example, serve with a creamy curry sauce made from the stock. Or you could bake a chicken pie with an almond flour crust (see the recipe on p. 78).

- 3 carrots
- 1 yellow onion
- 2 cloves garlic
- 1 8-inch (20-cm) leek
- 1 chicken, about 3.3 lbs. (1 ½ kg), preferably organic
- 2 bay leaves
- a few sprigs thyme and parsley
- 10 white peppercorns
- 1–2 tablespoons salt

Peel and cut the carrots, onion, garlic, and leek into small chunks. Rinse the chicken thoroughly—both inside and out—with cold water. Place the chicken in a large pot, and fill with water to cover the chicken. Bring the water to a boil, then pour off the hot water and fill the pot up with new water. This makes the stock clearer.

Bring the fresh water to a boil and, using a slotted spoon or ladle, skim off all the foam that bubbles to the surface. Now add in the carrots, onion, leek, garlic, herbs, and spices.

As I make my stock with a whole chicken that will become part of my dinner, I add salt or the chicken will be flavorless. Bring to a boil and let simmer under cover. Skim the foam occasionally. Place the peppercorns in a tea ball to make it easier to skim the surface without accidentally removing the peppercorns.

Remove the chicken from the pot after 45 minutes. Let it cool and pick the meat off the bones. If this is a cooking hen, it will need 2 hours in the pot to become tender. Place the carcass back into the stock and let it simmer, over low heat and without lid, for another hour or until the volume of water is reduced by half; it is now a concentrated stock. Strain the stock through cheesecloth into a bowl. Let cool.

The stock can keep for up to a week in the refrigerator, using the fat as a preserving layer on top. Or freeze the stock in smaller portions, to use as needed when flavoring soups and sauces. The stock can also be frozen in an ice cube tray, which makes it easy to use only what is needed at any particular moment.

You can also skip the reducing part and simply keep the lighter stock, which leaves you with a good, filling, and warming soup or drink.

Veal Stock

A typical sight in the restaurant kitchen where I worked was a pot of simmering veal stock on the stove. This became the base and flavoring for many a delicious sauce and soup.

I usually keep a bag of tendons, sinews, and meat scraps in the freezer, at the ready for stock.

- 2 large carrots
- 1 large yellow onion
- 1 8-inch (20-cm) leek
- 3.3–4.4 lbs. (1.5–2 kg) veal bones, chopped into chunks
- miscellaneous meat scraps
- 10 white peppercorns
- 3 bay leaves
- a few sprigs of rosemary, thyme, and parsley
- 1–2 tablespoons salt
- 3 tablespoons tomato purée
- water to cover the meat

Preheat the oven to 430°F (225°C). Peel and cut the vegetables into chunks. Place them with the bones and the rest of the ingredients—except the water—in a greased roasting pan. Add dollops of tomato purée on top so the purée will roast properly, as this mellows the acidity of the tomato a little. Roast the bones thoroughly for 40 minutes, until they have an appetizing color. Remove the pan after half the cooking time, and mix the bones so they get browned evenly all around.

Using a slotted spoon, set everything in a large stew pot. Pour water over until just barely covering the bones. Add salt, starting with the smallest amount suggested. Bring to a boil and let simmer, over low heat and without a lid, for 8 hours. Add more water occasionally, as it will evaporate as it boils.

Remove the bones from the pot and strain the stock into another pot. Let this stock cook without a lid until reduced to half its original volume. Taste for salt, and add more if needed. Strain again, and let it cool. The difference between a light stock and a concentrated stock is simply the length of the cooking time, to reduce the volume of stock to a dark stock; the flavor of a concentrated stock is far more intense.

Skim off the fat and fill an ice cube tray with the stock. Freeze the tray so you have the cubes on hand when you're making a soup or a sauce. This way you'll have a stock that's free of additives, which is something you can't get when buying stock in bottles at the grocery store.

Removing the fat from the stock before freezing it doesn't mean that I prefer a lean stock. I do it only because the fat oxidizes when it boils for so long, and as a result it doesn't taste very good.

Veal stock is usually prepared without salt, but I add some because I prefer the taste. I like to take a frozen cube of veal stock and add it to a mug of boiling water. With the addition of a tablespoon of organic coconut oil and some chopped herbs, this drink becomes a real powerhouse.

❧

Fish Stock/Fumé

It's both easy and quick to make homemade fish stock. The best stock is made from the bones of white-fleshed fish, but I'll use whatever I have on hand at the moment.

Use the stock for seasoning soups and sauces. The flavor and nutrition is far superior to any made from a stock cube or out of a bottle.

- about 2.2 lbs. (1 kg) fish bones and heads
- 3 tablespoon butter
- 1 carrot
- 1 4-inch (10-cm) leek
- 1 yellow onion
- 5 whole allspice berries
- 5 white peppercorns
- 3 star anise
- 1 bay leaf
- a few sprigs dill and parsley
- about 1 tablespoon salt
- cold water, enough to barely cover the bones

Chop the fish bones into coarse chunks, remove the gills from the heads, and rinse the bones thoroughly in cold running water. Peel and cut the carrot, leek, and onion into small chunks.

Brown the fish bones and vegetables lightly in a large pot. Let them fry for a bit, stirring often to prevent the mix from browning too much. Pour water into the pot and bring to a boil; skim the stock thoroughly. Add the spices, the sprigs of herbs, and salt, and let simmer for 20 minutes.

Using a strainer, pour off the stock from the bones and vegetables into another saucepan. If the bouillon is very cloudy or murky looking, strain it again, this time through a coffee filter or a piece of cheesecloth. Cook to reduce the stock to half its original volume to achieve a stronger flavored stock.

Freeze the stock in small jars or an ice cube tray, and use the cubes as needed for seasoning. Pick off the meat that's still attached to the fish bones, and use it in a soup or gratin.

BREAD

Walnut Bread

This bread is rich in walnuts, which imparts it with both a lovely taste and chewy texture. I mixed in apple cider vinegar to give the bread a slight tang, almost like a sourdough.

It's delicious when toasted, too. Try topping a slice with a hearty smear of butter and some goat cheese or salami.

- 5 ¼ oz. (150 g) walnuts
- 3 ½ oz. (100 g) coconut oil
- 2–3 tablespoons bread spices (ground anise, caraway seeds, fennel)
- 6 large eggs
- 8 ¾ oz. (250 g) coconut milk
- 2 tablespoons apple cider vinegar
- 6 ¾ fl. oz. (200 ml) almond flour
- 2 teaspoons baking powder
- about 1–1 ½ teaspoon salt
- ⅓ cup + 1 ½ tablespoons (100 ml) psyllium husk powder
- 3 ½ tablespoons (50 ml) flax seeds

Preheat the oven to 300°F (150°C). Grease a loaf tin (1 ½ quarts or 1 ½ liters) thoroughly.

Crush the walnuts coarsely with a mortar and pestle, or chop them with a knife. Warm the coconut oil together with the spices, and let it cool. In a bowl, mix the nuts with all the dry ingredients and start out with a small amount of salt. With a handheld electric mixer, whisk the eggs until thick, and then mix in the coconut milk, vinegar and coconut oil. Add the dry ingredients in with the mixer on low speed. Mix well and taste for salt and spices.

Use a rubber spatula to pour the dough into the loaf pan, and press on the dough's surface to make it level. Wet your hand in cold water and smooth out the surface. Cover the pan with aluminum foil and bake in the lower part of the oven for 45 minutes.

Remove the foil and bake the dough for another 45 minutes. Let the bread cool a little in the pan, and then unmold it on to a rack. Let it cool on the rack without covering it so it stays crusty. Slice the bread once it has cooled completely. Freeze the slices individually in order to enjoy them as needed; when defrosted in a toaster, the bread tastes freshly baked.

Halloumi Cheese Bread

If you like cheese and seeds, do I have bread for you! I only use egg whites here to avoid ending up with bread that tastes like omelet, which tends to be a problem with LCHF breads.

We use a lot of egg yolks in other dishes—to make mayonnaise, béarnaise, and hollandaise sauces—so there are always leftover egg whites in the refrigerator.

The Halloumi adds a good, chewy texture to these little rolls. They're tasty with just a bit of butter, and they pair very well with soup.

- 3 ½ oz. (100 g) Halloumi cheese
- ⅔ cup (150 ml) sesame seeds
- 3 ½ tablespoons (50 ml) flax seeds
- 2 tablespoon cold-pressed olive oil
- about ¼ teaspoon salt
- 3 large egg whites

Preheat the oven to 350°F (175°C). Line a baking sheet with a sheet of parchment paper.

Grate the cheese coarsely. In a bowl, mix the cheese with seeds, olive oil, and salt. As the cheese is rather salty, start the recipe off by using only half the amount of salt, and taste and adjust for seasoning as you go along.

Lightly whisk the egg whites and mix them with the cheese. Let the batter sit for 1 hour.

With two spoons, shape oval buns from the batter and place them on the prepared baking sheet.

This batter makes approximately 8 small rolls. Bake them at 350°F (175°C) for 15 to 20 minutes, or until golden brown. Let them cool on a rack. Freeze leftovers, and reheat them in the oven after they've been defrosted.

Seed Crisp Bread

A Facebook friend asked me if I could rework a recipe for seed bread that contained corn flour, wheat flour, and canola oil into a recipe that would work for those of us who eat according to LCHF/Paleo guidelines.

Here is the result: seed crisp bread—it has become very popular with both food bloggers and cookbook authors. It is also happens to be a hit with people who eat ordinary bread too.

A similar type of crisp bread is now being sold in Sweden under the name Spröda Frökex (Tender Seed Crackers). I worked with the bakery Huså Bröd (Huså Breads) to produce this product.

- ⅓ cup + 1 ½ tablespoons (100 ml) sunflower seeds
- ⅓ cup + 1 ½ tablespoons (100 ml) sesame seeds
- ⅓ cup + 1 ½ tablespoons (100 ml) whole flax seeds
- 1 tablespoon psyllium husk powder
- 1 tablespoon ground spices (anise, fennel, caraway seeds)
- about 1 teaspoon salt
- 1 cup + 1 tablespoon (250 ml) boiling water
- 1 ¾ fl. oz. (50 ml) coconut oil (unflavored)

Preheat the oven to 300°F (150°C). Line a cookie sheet, 12 x 13 inches (30 x 33 cm) with parchment paper. Mix all dry ingredients in a bowl. Bring the water and coconut oil to a boil. Pour the water/oil mixture over the dry ingredients and mix thoroughly. Taste for salt.

Pour the dough immediately onto the prepared baking sheet, using a rubber spatula to spread the dough. Place another sheet of parchment paper on top of the dough, and with a rolling pin roll the dough out to a thin layer—that's the easiest way to make it even.

Bake the dough at 300°F (150°C) for 45 minutes. Remove the baking sheet from the oven and, with a dough scraper or a sharp knife, mark the dough into long segments. Lower the oven's temperature to 122°F (50°C), and leave the bread in the oven to dry for 2 hours. Halfway through, break the dried dough into long pieces and turn them over.

Here I've chosen to flavor the bread with bread spices. If you'd rather not add them, you can sprinkle some salt flakes over the bread before it has finished baking, or you can flavor it with dried herbs instead.

Dairy-Free Savory Muffins

Many choose to omit milk from their daily diet due to sensitivity to dairy or other dietary guidelines. Here I present to you satisfying yet totally dairy-free muffins. Their texture will remind you of that of scones.

Black cumin imparts a very nice flavor to the muffins. In Latin, the name of black cumin is Nigella sativa; cumin seeds are similar to kalonji seeds, and can be found in any spice shop.

Makes 6 to 8 muffins

- 5 large eggs
- 1 tablespoon psyllium husk powder
- 6 ¾ fl. oz. (200 ml) almond flour
- ¼ cup (50 ml) coconut flour
- 2 teaspoon baking powder
- ½ teaspoon salt
- 1 ¾ fl. oz. (50 ml) coconut oil

To Sprinkle on Top

- ½ teaspoon black cumin or kalonji seeds
- ½ teaspoon sesame seeds

Preheat the oven to 400°F (200°C).

Mix all dry ingredients in a bowl. With a handheld electric mixer, whisk the eggs until very thick—this will take several minutes. With a rubber spatula, fold in the dry mixture, alternating with the coconut oil. Mix thoroughly. Leave the batter to rise for a few minutes. With two tablespoons, spread the batter into baking cups (doubled up, by nesting one cup inside another), or in muffin tins placed on a baking sheet. Sprinkle the batter with some black cumin or kalonji seeds, and sesame seeds.

Bake the muffins in the middle of the oven for 12 to 15 minutes. Let them cool on a baking rack. Freeze any muffins not eaten right away, and defrost them as needed. Freshen them by putting them in a warm, 300°F (150°C) oven for a few minutes once they're defrosted.

SAUCES AND TASTY GREEN SIDE DISHES

Herb Aioli

Olive oil provides a characteristic taste to aioli. By crushing the herbs with a mortar and pestle, you'll release a lot of their natural oils into the sauce, which will give it an extra shot of flavor, color, and nutritional value.

For variety, try flavoring the aioli with a pinch of ground saffron instead of using herbs.

- 3 cloves garlic
- ¼–½ teaspoon salt
- 3 large egg yolks (from free-range eggs)
- 1 ¼ cups (300 ml) mild, cold-pressed olive oil
- 2–3 tablespoons finely chopped herbs
- juice from ½ lemon

Crush the garlic; mix it with the smaller amount of salt and the egg yolks in a tall, narrow bowl. Add the oil, at first drop by drop and then in a thin stream, while whisking vigorously. This is easily done with an immersion blender, but a handheld electric mixer will work, too.

Crush the chopped herbs in the mortar along with the lemon juice and a small pinch of salt until the herbs are completely mashed; stir into the aioli.

Taste and adjust for salt—you should aim for a pleasant combination of tart and salt.

Chill the aioli for a few hours to allow the flavors to develop.

Mayonnaise

Mayonnaise is one of the most basic yet versatile condiments for anyone who eats food low in carbohydrates, as it can be incorporated into many different sauces and mixes. If you make your own mayonnaise using free-range eggs, it will be both better tasting and more nutritious than anything sold in stores.

I use mild-tasting olive oil, and not cold-pressed, as its flavor is too assertive for mayonnaise.

- 2 large egg yolks, from free-range eggs
- 1 ½ teaspoons unsweetened mustard
- 1 teaspoon white balsamic vinegar
- ⅛–¼ teaspoon salt
- 6 ¾ fl. oz. (200 ml) mild-flavored olive oil
- some freshly ground white pepper
- a few drops of Worcestershire sauce

Mix the egg yolks, mustard, vinegar, and salt with an immersion blender in a narrow, tall vessel. Add the oil carefully, first drop by drop and then in a thin stream. Blend continuously. You can add the oil in larger amounts once the mayonnaise has started to emulsify. Work the blender up and down the vessel to ensure everything mixes thoroughly.

Season with white pepper and Worcestershire sauce. Taste and adjust for vinegar or salt. Put the mayonnaise in a jar with a tight-fitting lid. If you prefer mayonnaise with more zing, double the amount of mustard.

Haydari

Haydari, which I eat a lot of during my winter months in Turkey, is a tasty alternative to mayonnaise. This garlic-flavored yogurt is found on meze buffets, and you can also buy it in grocery stores. The inclusion of mint adds a fresh touch that makes it pair exceptionally well with lamb.

A dollop of haydari is also lovely in a bowl of goulash, as a side to fried fish, or as a dip for vegetable sticks. If you'd like a creamier sauce, use smetana or crème fraîche instead of yogurt.

- 6 ¾ fl. oz. (200 ml) Turkish (or Greek) yogurt
- 2 tablespoons cold-pressed olive oil
- 2 tablespoons fresh, chopped mint or 1 tablespoon dried mint
- 1 tablespoon finely chopped Italian parsley
- 1 garlic clove, crushed
- about ½ teaspoon sea salt

Whisk the yogurt and oil together thoroughly. Mix in the chopped herbs and the garlic; season with salt. Chill the sauce for a few hours to allow the flavors to develop.

Taste and adjust for mint and salt.

Whitefish Roe Sauce

This dish of whitefish roe is a perennial favorite at my sister's buffet table. It goes well with smoked and cured salmon, paté, fish cakes, and shellfish. According to my eldest niece, it even goes well with meatballs!

It's fine to use other whitefish or salmon roe. They contain no additives, which is not the case with all of the fish roe that is sold under the name caviar.

- 2 tablespoons mayonnaise (see recipe on p. 124)
- 6 ¾ fl. oz. (200 ml) smetana (or quark)
- 2 tablespoons finely chopped dill
- juice from 1 lemon
- 1 crushed garlic clove
- about ⅛ teaspoon salt
- about ⅛ teaspoon white pepper
- 1 ¾ oz. (50 g) whitefish roe

Mix all the ingredients—except the fish roe—in a bowl. Carefully fold in the roe, and taste to adjust for salt or lemon.

Chill the sauce for an hour before serving to let the flavors bloom fully.

Marinated Feta Cheese

Marinated goat's or sheep's cheese turns into a delicious side dish when paired with pulled pork or tapas, or when added to a satisfying salad.

You can also marinate other cheeses such as cubes of aged Emmenthaler or Cheddar.

- 10 ½ oz. (300 g) feta cheese, from either sheep's or goat's milk
- 3 cloves garlic
- peel of 1 organic lemon
- ½ tablespoon dried thyme
- ½ tablespoon dried oregano (Mediterranean)
- 2 tablespoons pink peppercorns
- cold-pressed olive oil

Cut the feta in half, and then into smaller rectangular chunks. Peel and slice the cloves of garlic. Wash and grate the lemon peel with a zester. Place the cheese in a glass jar and layer with the herbs, garlic, lemon zest, and pink peppercorns. Add in the olive oil to cover the cheese and seasonings completely—the cheese must always be covered by oil to prevent oxygen from entering the jar.

This way, the cheese will keep for several weeks at room temperature. The flavors are stronger if the cheese is stored at room temperature than if kept in the refrigerator.

Eggplant Salad

Grilled eggplant and bell peppers are often on the menu during my winter stays in Turkey, where vegetables are typically cooked on a coal grill in the garden in the evening. The aromas wafting around the neighborhood are divine!

Here at home, it's easy to grill vegetables in the oven until their skins are charred and black, which imparts a slightly smoky flavor to the flesh inside. These vegetables contain very few carbohydrates, making them a perfect side dish to any LCHF/Paleo meal.

- 2 eggplants
- 2 green bell peppers
- 1 red chili pepper
- 3 ⅓ fl. oz. (100 ml) sprigs of parsley
- 2 cloves garlic
- juice from 1 lemon
- about 1 teaspoon salt
- 3 ⅓ fl. oz. (100 ml) mayonnaise (see recipe on p. 124)

Preheat the oven to 480°F (250°C).

Cut the eggplants and bell peppers in half lengthwise, and seed the bell peppers. Place the vegetables peel side up in an ovenproof dish. Set the dish in the middle of the oven and bake for half an hour, turning the eggplants after 20 minutes. Remove the dish from the oven and let the vegetables cool down a little.

Cut the chili in half and seed it. Use rubber gloves to complete this task, and do not touch your face! For a spicier eggplant salad, leave the seeds in. Chop the chili very finely. Chop the parsley and crush the garlic. In a bowl, mix it all with the lemon juice and salt. Pull the skins off the grilled vegetables and cut them into small cubes; add to the parsley mix and the mayonnaise. Mix thoroughly. Chill for a few hours to give the flavors time to bloom.

Feta Cheese Sauce with Brazil Nuts

Smetana is excellent for making cold sauces—it has a milder tang than crème fraîche and is richer in fat. Brazil nuts are also very high in natural oils, and contain very few carbohydrates—only about 3.4 percent. However, their nutrient count is very high, especially in selenium, which is often not easy to get enough of.

Consequently, this sauce is a very tasty and nutritious food. It makes a great topping on a fish or vegetable gratin.

- 3 ½ oz. (100 g) feta cheese
- ⅓ cup + 1 ½ tablespoons (100 ml) smetana (or quark)
- ⅓ cup + 1 ½ tablespoons (100 ml) coarsely chopped Brazil nuts
- 2 teaspoons lemon juice
- 1 tablespoon cold-pressed olive oil
- 1 tablespoon finely chopped Italian parsley
- salt, if needed

Crumble the feta cheese into a bowl and mash it with a fork. Mix it thoroughly with the smetana and the nuts. Stir in the lemon juice, olive oil, and parsley. Taste and adjust for salt, if needed.

Chill the sauce for a few hours to let the flavors blend, and the sauce will become even tastier.

Tips for Building Satisfying Salads

A fresh salad can kick a meal up a notch or two. While it tastes good, we also get lots of vitamins and minerals, especially if we add more salad ingredients than iceberg lettuce, tomato, and cucumber.

You can find tender lettuce leaves sold by weight in many grocery stores today, so go ahead and try a variety of types. If you buy whole heads of lettuce, try grating them instead of slicing them into strips with a knife. The salad will taste and look a lot better.

Here are some suggestions for different salad combinations:

- Swiss chard, Italian parsley, crumbled goat's cheese, sliced radishes, and pine nuts. Dress with cold-pressed olive oil and lemon juice.
- Arugula, spinach leaves, crumbled blue cheese, grilled bell pepper, walnuts, and thyme. Dress with mild olive oil and dark balsamic vinegar.
- Radicchio, frisée lettuce, mâche lettuce, julienned leek, pomegranate seeds, pumpkin seeds, and coarsely chopped dill. Make a dressing from mayonnaise, lemon zest, lemon juice, salt, and some water.
- Lollo rosso lettuce, green bell pepper, asparagus, hazelnuts (filberts), chives, topped with shaved Västerbotten (or Parmesan) cheese. The dressing is made with mild olive oil, apple cider vinegar, Dijon mustard, some salt, and coarsely ground black pepper.
- Romaine lettuce, hard-boiled eggs, avocado, cherry tomatoes, hothouse cucumber, parsley, and chives. Top with avocado oil with lemon and some salt flakes.
- Baby spinach leaves, mâche lettuce, white asparagus spears, shaved Parmesan, lemon thyme, roasted sesame seeds, and pomegranate seeds. Dress with cold-pressed olive oil, white balsamic vinegar, some pressed garlic, and salt flakes.
- Oak leaf lettuce, parsley, lemon balm, diced hothouse cucumber, kiwi, and roasted hazelnuts (filberts). Cold-pressed olive oil and apple cider vinegar make a nice dressing.
- Lollo rosso lettuce, mâche lettuce, coarsely chopped Brazil nuts, diced apple, avocado, and chives with lime juice, chili flakes, and mild olive oil for dressing.
- Radicchio, Swiss chard, finely cut red onion, halved cherry tomatoes, crumbled feta cheese, and thyme. Cold-pressed olive oil and red wine vinegar make a dressing.

❧❧

Fennel Slaw

One of the most common and nutrient-packed salads is coleslaw made with white cabbage and carrots. My juicy fennel slaw is slightly different: its mild spice comes from the light anise flavor of fennel and peppery arugula.

It becomes even more flavorful when paired with an herb- and garlic-infused aioli.

This salad is excellent as a side to fried fish or chicken.

- 1 ⅔ cups (400 ml) finely shredded fennel
- 6 ¾ fl. oz. (200 ml) finely shredded arugula
- ⅓ cup + 1 ½ tablespoons (100 ml) julienned leek
- ⅓ cup + 1 ½ tablespoons (100 ml) herb aioli (see p. 122)
- about ¼ teaspoon salt
- about ½–1 teaspoon red chili flakes

Mix all the shredded vegetables in a bowl. Fold in the aioli and season with salt and chili flakes. Refrigerate for a few hours to let the fennel soften and the flavors develop.

Red Cabbage Salad with Horseradish

This tangy salad is a good alternative to beet salad that's low in carbohydrates. Why not include it in on your Christmas or Easter buffet along with a variety of cold cuts?

The salad also works very well instead of traditional beets to accompany veal brawn and hash browns.

- 2 cups (500 ml) red cabbage, finely shredded
- 6 ¾ fl. oz. (200 ml) mayonnaise (see recipe on p. 124)
- ¼ cup (50 ml) grated fresh horseradish
- juice from ½ lemon
- 1 tablespoon apple cider vinegar
- about ¼ teaspoon salt
- about ⅛ teaspoon pepper

Put the shredded cabbage in a bowl. Squeeze it vigorously with your fingers to make the salad really juicy. Stir in the mayonnaise, horseradish, lemon juice, and vinegar; mix thoroughly. Season with salt and pepper; if you prefer a salad with a bit more bite, add some more horseradish to taste.

Keep chilled for a few hours, or preferably until the next day—the salad will taste even better.

Savoy Cabbage Simmered in Cream with Hot-Smoked Pork Belly

In the past, diabetics and other people who needed to keep their blood sugars under control were encouraged to eat side dishes such as white cabbage simmered in cream.

My LCHF version of this dish is made with savoy cabbage and hot-smoked pork belly. You can trade the savoy cabbage for kale, which is reminiscent of the Swedish province of Halland's traditional Christmas cabbage dish, which happens to be stewed kale flavored with sugar.

- 1 ¾ oz. (50 g) hot-smoked pork belly
- 1 ¾ oz. (50 g) butter
- 4 ¼ cups (1000 ml) shredded savoy cabbage
- about ¼ teaspoon white pepper
- ¼ teaspoon nutmeg
- 1 ¾ fl. oz. (50 ml) veal stock (see p. 106) OR ¼ cube organic meat stock and 1 ¾ fl. oz. (50 ml) water
- 6 ¾ fl. oz. (200 ml) heavy cream
- about ¼ teaspoon salt

Cut the pork belly into thin slices, then finely julienne those slices. Cook the pork belly strips in the butter until lightly brown. Add the cabbage and season with pepper and nutmeg, and sauté for about 10 minutes, stirring constantly.

Add the stock or diluted bouillon cube, and let the cabbage leaves cook and absorb the stock, which will give them a wonderful flavor. Pour in the cream, and let the cabbage simmer for about 30 minutes, until it turns into a thick stew. Stir occasionally. Taste and adjust for salt (although both pork belly and stock already contain some salt).

This way of preparing cabbage makes it a good side for several dishes, such as chicken, fish, meatballs, and good quality sausages.

Experiment by serving the cabbage with fried fish, as it makes a very tasty combination.

Fennel and Coconut Gratin

Vegetable gratins fit very well into menus emphasizing fewer carbohydrates and more natural fats. The following pages feature three dishes that are excellent sides to many of the slow-cook recipes, and all three are quick and easy to make.

The flavors of fennel and coconut pair very well together. Here they are simmered in a creamy gratin redolent of coconut cream and lime. This dairy-free dish goes very well with chicken or with hot-smoked salmon or other fatty fish.

- 2 fennel bulbs
- 2 tablespoons coconut oil (with intense coconut flavor)
- ⅓ cup + 1 ½ tablespoons (100 ml) leek, julienned
- ⅓ cup + 1 ½ tablespoons (100 ml) toasted, unsweetened coconut flakes
- about 1 teaspoon salt
- about ¼ teaspoon red chili flakes
- juice from 1 lime
- 8 ¾ oz. (250 g) coconut cream

Preheat the oven to 435°F (225°C). Grease an ovenproof dish with a small amount of coconut oil.

Cut the fennel into quarters, and cut away the tough stem. Slice each quarter into strips. Heat the coconut oil and sauté the fennel, leek, and coconut flakes for one minute without letting them brown; season with salt, chili, and lime juice. Add in coconut cream and bring to a boil. Transfer it all into the prepared ovenproof dish. Bake at 435°F (225°C) for 10 minutes. For variety, vary the seasoning by adding in some saffron or curry.

Bell Pepper and Chèvre Gratin

Chèvre, a French goat cheese, is delicious when baked with bell peppers. The sweetness from the bell peppers makes a nice counterpoint to the saltiness of the cheese. A splash of vinegar over the vegetables enhances the flavors even further.

- 1 red bell pepper
- 1 orange bell pepper
- 1 green bell pepper
- 1 red onion
- 1 oz. (25 g) butter
- 3 ½ oz. (100 g) chèvre
- about ½–1 teaspoon salt
- about ¼ teaspoon white pepper
- 1 tablespoon apple cider vinegar

Preheat the oven to 435°F (225°C).

Cut the bell peppers in even-sized strips, and the onion in narrow wedges. Lightly brown the butter, and sauté the bell peppers and onion a few minutes; season with salt and pepper. Add the vinegar and stir so everything is thoroughly mixed. Transfer to an ovenproof dish, and top with slices of chèvre. Bake for 5 minutes.

Brussels Sprouts Baked in Blue Cheese

The Brussels sprout is a vegetable that didn't come into my life until my later years. I didn't appreciate its qualities until I began cooking dishes containing fewer carbohydrates. My intake of vegetables has increased significantly since the days when I mainly filled up on pasta.

- 2 cups (500 ml) Brussels sprouts
- 6 ¾ fl. oz. (200 ml) smetana, crème fraîche, or quark
- 6 ¾ fl. oz. (200 ml) creamy blue cheese, grated
- about ¼ teaspoon salt
- about ⅛ teaspoon white pepper
- a small pinch of grated nutmeg

Preheat the oven to 400°F (200°C). Butter an ovenproof dish.

Parboil the Brussels sprouts for 3 minutes in salted water. Drain them well in a colander. Cut the sprouts in half lengthwise and place them in a bowl. Mix in the smetana and the grated blue cheese. Season and mix well. Pour into the prepared baking dish and bake for 20 minutes.

The creamy saltiness of this dish goes very well with fatty fish, pork, or chicken.

Blue Cheese Butter

A seasoned butter provides additional flavor and enhances many dishes while also making them more nutritious. I always eat extra butter with my food, and it keeps me full and satiated a long time. I hardly remember when I felt the need to eat a snack between meals.

Prepare a large batch of herb butter so you'll always have some to go with your food.

Leave the butter at room temperature for a good while if it is to be combined with other ingredients.

- 1 ¼ lbs. (500 g) butter
- 3 ½ oz. (100 g) creamy blue cheese
- ⅓ cup + 1 ½ tablespoons (100 ml) finely chopped chives
- salt and pepper to taste

Put the butter in a bowl and crumble in the blue cheese. Add in the chives; using a hand-held electric mixer, whip the butter until it is light and airy and everything is thoroughly mixed. Fill some small jars, pipe little roses, or roll up the butter into the shape of a log. Store the butter in the refrigerator or freezer.

Red Onion Butter

- 1 red onion
- 3 ⅓ fl. oz. (100 ml) dry red wine
- about ¼ teaspoon salt
- chili flakes, to taste
- 1 ¼ lbs. (500 g) butter
- ¼ teaspoon Worcestershire sauce

Chop the red onion into fine cubes, and place in a saucepan. Add red wine, salt, and chili flakes. Bring to a boil and let simmer until the onion has absorbed all the wine. Let cool. Whip the onion with the butter and season with Worcestershire sauce. Taste and adjust for more salt or chili if needed. Roll up the butter into the shape of a log, or fill some small individual bowls with the butter.

Mustard Butter

- 1 ¼ lbs. (500 g) butter
- 2 tablespoons coarse ground Dijon mustard
- 1 tablespoon smooth Dijon mustard
- salt and pepper to taste

Whip butter and mustards until white and fluffy. Add seasoning if necessary. Pipe roses of the butter onto parchment paper and freeze, roll into the shape of a log, or fashion the butter into small eggs with a spoon dipped in warm water.

How to Roll Butter into a Log:

On a sheet of parchment paper, set the butter down along the midline. Fold one side of the paper over. Run a broad-bladed knife over the paper toward the butter, until you make an even cylinder. Roll the butter in the parchment paper like a Swiss roll, and twist the paper at each end of the log like a candy wrapper. Keep the log of butter in the refrigerator or freezer. Bring the log to room temperature before serving it to let the flavors develop fully.

Avocado Butter

"Butter" made from avocado and coconut oil is a great alternative to regular butter if you follow a dairy-free diet. It's also a very tasty spread on one of my dairy-free low-carbohydrate breads—refer to the seed crisp bread (recipe on p. 116) or the savory muffins (recipe on p. 118).

This avocado butter is also a good substitute for herb butter with any of our dishes. Make sure to season it first, preferably with some garlic, fresh herbs, chili, and some lemon or vinegar.

- 2 ripe avocados
- 2 tablespoons coconut oil (unflavored)
- 1 tablespoon cold-pressed olive oil
- ¼–½ teaspoon salt

In a bowl, mash the avocado with a fork. With a handheld electric mixer, whip in the oils until creamy. Or, mix them directly in the bowl with an immersion blender; season with salt. Chill for an hour or so to let the butter firm up and the flavors blend completely.

Avocado butter can keep in the refrigerator for a few days if you leave the avocado stone in the bowl to prevent the butter from turning brown. This butter can also be frozen in small portions. Bring it back to room temperature for a while before serving it to allow the oils to soften a bit.

Recipe Index

Notes